from Conception to Birth: a Life Unfolds

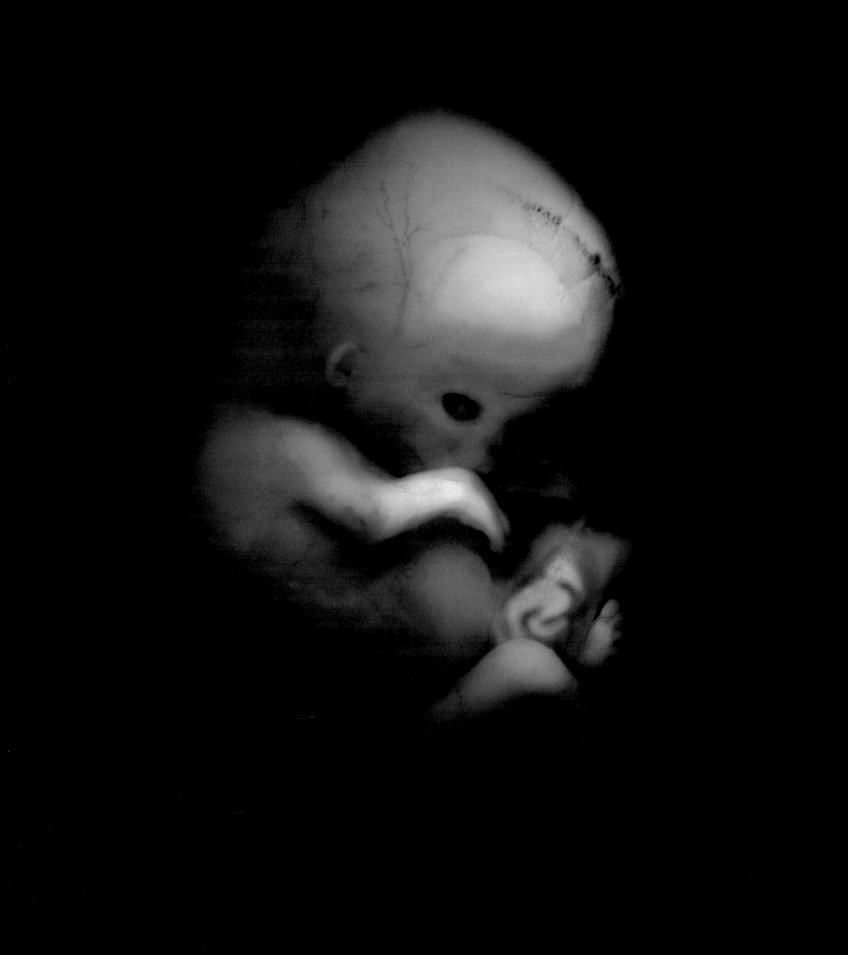

what's going on with the baby now?

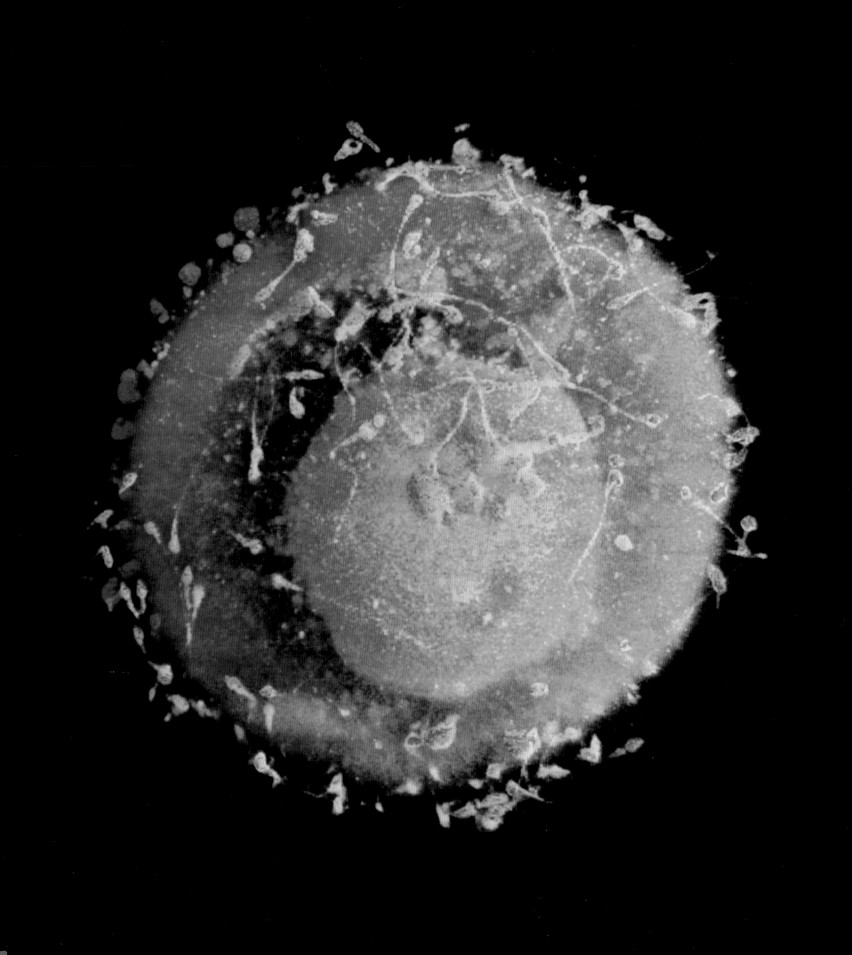

celebrate

Our living architecture... stirring, beautiful, wondrous—

in every sense of the word *marvelous.*

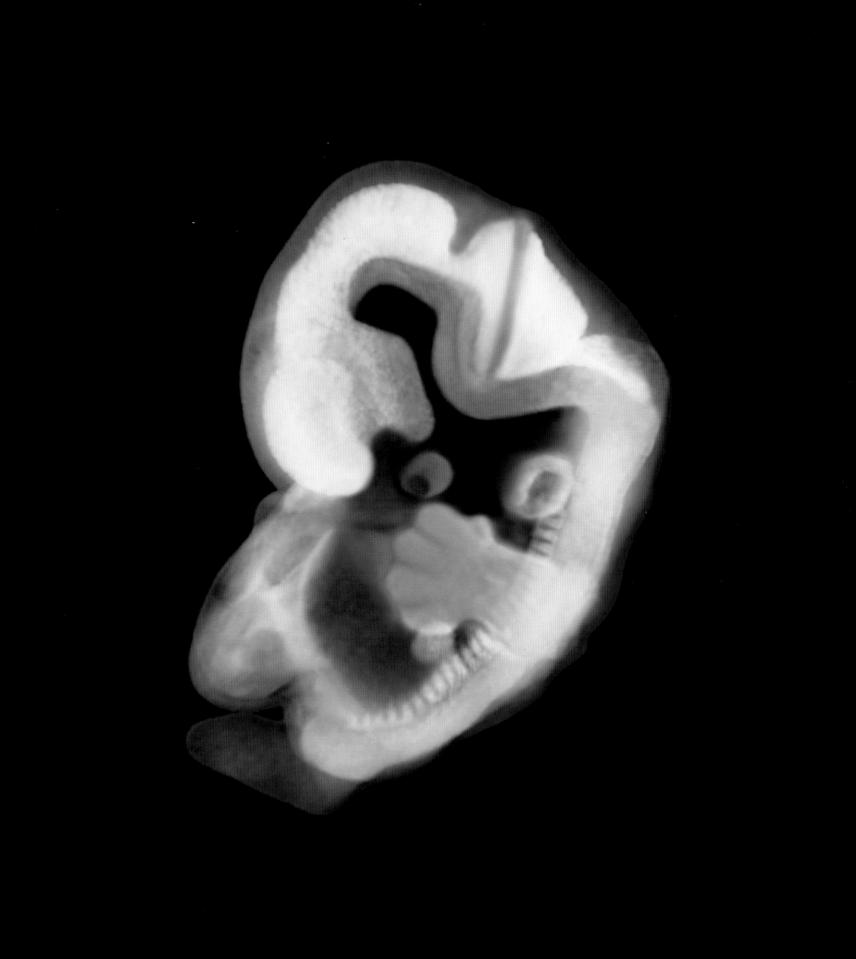

"...from so simple a beginning, endless forms most beautiful and most wonderful have been, and are being evolved."

— *The Origin of Species,* Charles Darwin

Alexander Tsiaras

Text by Barry Werth

Doubleday

New York   London   Toronto   Sydney   Auckland

from Conception to Birth: a Life Unfolds

Published by Doubleday, a division of Random House, Inc.

1540 Broadway, New York, New York 10036

DOUBLEDAY and the portrayal of an anchor with a dolphin are trademarks
of Doubleday, a division of Random House, Inc.

ISBN 0-385-50318-0

Copyright © 2002 by Alexander Tsiaras

Library of Congress Cataloging-in-Publication Data

Werth, Barry

    From conception to birth: a life unfolds/Alexander Tsiaras;

    text by Barry Werth.--1st ed.

    p. cm.

    1. Prenatal diagnosis. 2. Diagnostic imaging. I. Tsiaras, Alexander. II. Title.

RG628.W47 2002

618.3'2075--dc21   2002024707

PRINTED IN JAPAN

November 2002

First Edition

1 2 3 4 5 6 7 8 9 10

From Conception to Birth is available at a discount when ordering 25 or
more copies for sales, promotions, or corporate use. Special editions,
including personalized covers, excerpts and corporate imprints, can be
created when purchasing in large quantities. For more information
please call (800) 800-3246 or e-mail specialmarkets@randomhouse.com.

Design: Stark Design•NYC: Adriane Stark and Craig Bailey

This work is dedicated to the individuals who have had the foresight to assemble and preserve the Carnegie Human Embryology Collections. The continuing commitment of resources from the National Institutes of Health, through the National Museum of Health and Medicine and the American Registry of Pathology, has made it possible for scientists and artists to advance our knowledge and our marvel of the process of human development.

We also wish to thank Adrianne Noe, Ph.D., Director of the National Museum of Health and Medicine of the Armed Forces Institute of Pathology, and the AFIP for their dedication to the collection and encouragement to me and others who wish to share their research and wonder of human development.

Contents

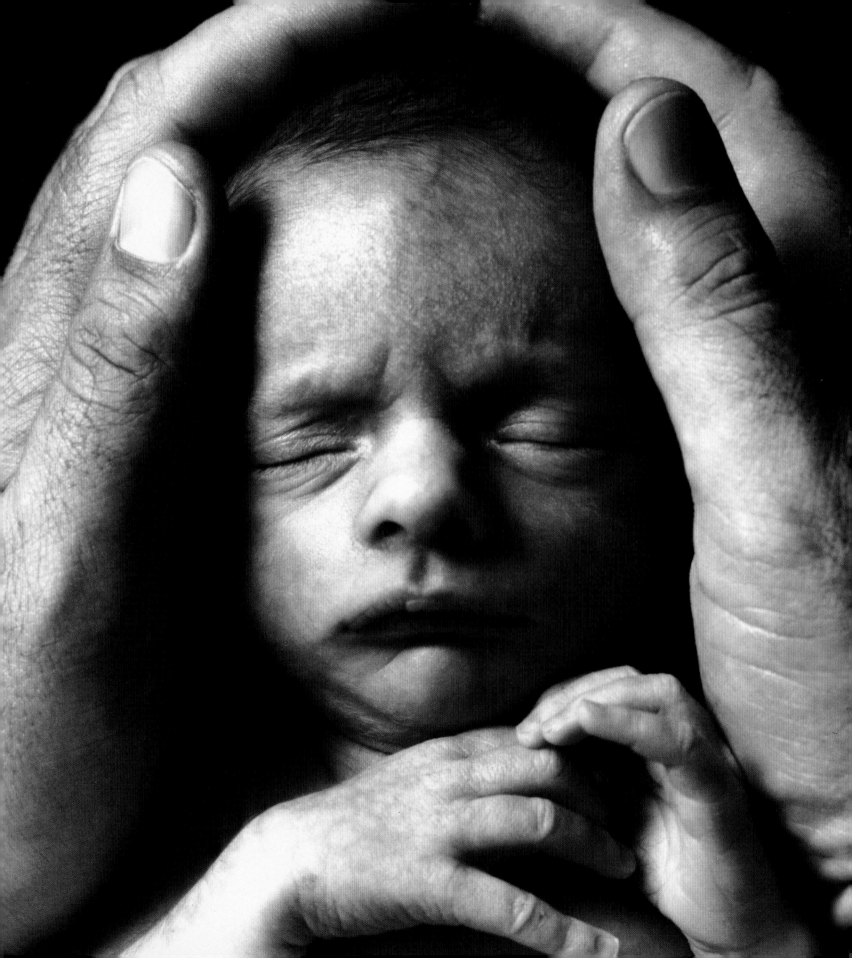

the Drama of Life Unfolding

What's happening with the baby now? When our grandparents and even our parents asked this question, the answer was locked in mystery, like the night sky. They knew that as a child grows and develops inside its mother's uterus, a new life unfolds. But they never anticipated they might someday observe this inner cosmos. Even if your mother and father had a rough idea about the stages of embryonic development from biology class or brochures in her obstetrician's office, they couldn't visualize the wondrous activities stirring within her as she became pregnant.

That all started to change in the 1960s. Cameras began going everywhere, including the inmost parts of a woman's body. Suddenly it was possible to view the stages of unborn life through grainy ultrasound images or marvelous color photos in magazines and on TV. If heavenly bodies no longer were so mysterious in the age of space travel and X-ray telescopes, neither was the creation in utero of human bodies in an era of electron microscopes and tiny endoscopes that could peer inside a woman's womb. As the philosopher and ethicist Meredith Michaels observed, the language of space exploration and human embryology inevitably collided. "The blastocyst has landed!" *Life* trumpeted.

These new images upended the way we look at our own beginnings, by taking expectant parents  and others inside the hidden world of procreation and giving them their first glimpses of what happens as cells fuse, then divide and subdivide, to form a new person. As the baby developed, more and more refined images revealed a profound and beautiful world — part Hubble telescope, part Jacques Cousteau — in which miniscule cosmic voyagers floated in saline seas, reflexively sucking their thumbs. It was thrilling, inspiring, as we searched for familiar signs, experiences, moments, not only in our children's formation but our own.

Yet the images we saw were biographies of the unborn self, not autobiographies — not the inner experience, viewed from within. As marvelous as our understanding of what was happening with the baby had become, what was really going on — the cell-by-cell, tissue-by-tissue sculpting of a human being — remained all but shuttered from view.

Until now. As the pictures in this book reveal, recent developments in science and technology have vastly enhanced our ability to gaze inside ourselves and witness the drama of early human life — as it unfolds, from conception to birth. As biologists have decoded the molecular basis of life, computer scientists have developed three-dimensional techniques for scanning and displaying the body, which can isolate systems (nervous, skeletal, circulatory, etc.) and allow us to view them down to a molecular level.

What's happening with the baby now?

A generation ago the answer might well have been: "Welcome to the dark side of the moon."

And now?

Feast your eyes.

IN THE BEGINNING

The first primer on the combined topics of sex, science, and reproduction (and thus a valuable forerunner of this book) was published in the United States nearly a century ago, in 1911. *A Child's Guide to Living Things* by Edwin Tenney Brewster was no mere children's book, however. A Harvard-trained zoologist, Brewster wrote widely on scientific matters, and he sought to provide young readers with a modern understanding of how life begins. Modesty and discretion prevented him from addressing the "facts" of life; he wrote instead about starfish eggs and sea urchins. And human embryology remained a primitive, largely forbidden field. But some biologists and doctors had recently begun to explore how cells reproduce and change at the earliest stages of life, and Brewster explained in simple terms what they had — and hadn't — learned:

> So we are not built like a cement or a wooden house, but like a brick one. We are made of little living bricks. When we grow it is when these little living bricks divide into half bricks, and grow into whole ones again. *But how they find out when and where to grow fast, and when and where to grow slowly, and when and where not to grow at all is precisely what nobody has yet made the smallest beginning to find out.* [italics added]

Now leap ahead 9 decades to a university laboratory in Wisconsin, as an obscure, hang-gliding biologist  named James Thomson coaxes cells from a 4-day-old human embryo to reproduce in a glass dish. Since the 1970s, scientists routinely have "immortalized" human cells by tricking them into dividing again and again outside the human body. But oh, what cells! Thomson's "little living bricks" — embryonic stem cells — possess a protean capacity to morph into the more than 200 other types of building blocks that, copied a million billion times and working in magnificent concert, make up a human being. Contained in these microscopic units are virtually all the data needed for answering Brewster's enduring questions — how do living things know when, where, and how to build themselves?

This was — and is — life's crowning wonder, and Brewster deserves belated credit for inviting his young readers to ponder it at a time when the most advanced tools for understanding basic biological mysteries were, in retrospect, primitive and crude, the microscope and the X ray. Yet he, too, couldn't help but marvel at how a perfect sphere of cells suddenly elongates, inverts on itself, stretches, bends, and molds into elaborate living shapes. Starting from the point of conception—the unseen mingling of 2 coils of chemical information—the transformation defies imagining. Witness the young starfish egg, he wrote:

> One would perhaps expect to see the oil and jelly mixture change gradually into a starfish. Instead of this, however, this little balloonlike affair splits squarely in 2, and makes 2 little balloons just alike, and which lie side by side. . . . In about half an hour, each of these balloons or bubbles, "cells" as they have come to be called, has divided again; so that now there are 4. The 4 soon become 8; the 8, 16. In the course of a few hours, there are hundreds, all sticking together and all very minute; so that the whole mass looks like the heap of soap bubbles which one blows by putting the pipe under the surface of the soapsuds. . . .

Of course, this is only the beginning, a warm-up act. The most spectacular metamorphoses come, Brewster noted, in the weeks that lay just ahead:

> If it is an animal like ourselves, this body stuff, before it becomes a body, is a round ball. [Then] a furrow doubles in along the place where the back is to be and becomes the spinal cord. A rod strings itself underneath this, and becomes the backbone. The front end of the spinal cord grows faster than the rest and becomes larger, and is the brain. The brain buds out into the eyes. The outer surface of the body, not yet turned into skin, buds inward and makes the ear. Four outgrowths come down from the forehead to make the face. The limbs begin as shapeless knobs, and grow out slowly into arms and legs. . . .

Writing for children under age 10 in an era when biologists were better at describing things than explaining them, Brewster might be blamed for oversimplifying—little living bricks, soap bubbles, furrows, rods, knobs. And yet there is something elegant and prescient in each of his descriptions—he seemed to sense some divine infrastructure, some physical laws governing human construction, that modern biologists like Thomson have lately made only "the smallest beginning" to puzzle out.

In fact, as science has shown, we *are* like buildings, although the structures we most resemble are infinitely more majestic than the grandest brick house. A house is built from the ground up and assembled from different materials imported from someplace else. But people, like all living things, grow from the inside out, transforming as we go.

Imagine yourself as the world's tallest skyscraper, built in 9 months and germinating from a single brick. As that "seed" brick divides, it gives rise to every other type of material needed to construct and operate the finished tower — a million tons of steel, concrete, mortar, insulation, tile, wood, granite, solvents, carpet, cable, pipe, and glass as well as all furniture, phone systems, heating and cooling units, plumbing, electrical wiring, artwork, and computer networks, including software. This brick and its daughter bricks also know exactly how much of each to make, where to send them, and when and how to piece it all together. Now imagine further that when the building is done it has the capacity to love, hate, converse, do calculus, compose symphonies, and have rapturous physical relations with other towers, a prime result of which is to create new buildings even more elaborate than itself.

How does this happen? We know that like all else in the universe, not just other buildings, we are built of molecules, which consist of matter and energy but are not what any of us would call alive. *What directs our molecules to build?*

Science has begun to answer that question, too. Plainly one of the great biological discoveries between the publication of Brewster's *A Child's Guide* and when James Thomson was honored in August 2001 on the cover of *Time* is the understanding that what tells molecules 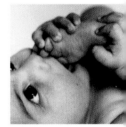 to make a human being are other molecules. All living cells are comprised mainly of 2 types of organic molecules — nucleic acids and proteins. Nucleic acids — genes — carry the instructions the cells need to function and reproduce; in other words the overall plans for the building, as well as for each cell's own small part. Proteins do the work, erecting scaffolding, chopping each other up, recombining, making more proteins. Indeed inside every cell teems a subuniverse — another skyscraper — comprised of tens of billions of proteins conducting complex chemical reactions at rates of ten billion times per second. Meanwhile, other molecules on the cell's surface interact with — talk to — molecules on the surface of other cells. Such, literally, is life, which is distinguished from nonlife by the profoundest activity of all: making self-copies. Reproduction.

"Molecules," as the embryologist Lewis Wolpert puts it, "are the natural language of the cell."

So we are like buildings, but buildings that replicate. And yet we're more than that. We're replication machines. We replicate because our cells replicate. They replicate because their molecules replicate. Everything about us seems to want to replicate. All the time. "The procreant urge of the world," Walt Whitman called it.

Which is why we love life and love children and love sex.

NATURE'S DESIGN

Like all great buildings and machines, the unborn self combines exquisite architecture and canny engineering.

Start with conception. Because it's necessary, when wanting a baby, to unite genes from a man and woman, and since these genes are encased in specialized cells inside the human body, it is essential (or was, until the advent of artificial fertilization) to steer two sexually mature bodies into intimate contact.

And so what happens? A meaningful look. A seductive Beaujolais. An inviting scent. Scientists studying sweat lately have isolated chemical secretions called pheromones that signal our sexual ripeness and availability. The point: because our genes need to combine, our egg and sperm cells need to fuse; and enhancing desire while loosening inhibitions is the surest way to improve the chances in humans that both these things will occur.

Throughout the next 9 months — indeed, throughout life — the tools and processes of procreation are similarly optimized. Getting male genes to female genes? No problem. A mature, healthy man produces several hundred million sperm cells a day, and while the force of ejaculation is generally enough to propel them just halfway toward their goal, each is also packed with a trail mix of rare supersweet sugar to fuel it the rest of the way. Penetrating the famed zona pellucida, the cunning matrix of sugar and protein surrounds the mature egg like a force field. Molecules on the surface of the sperm heads are designed to bind specifically to molecular "receptor sites" — portals — on the zona, which imbibes one sperm, then stiffens to repel all others. "The egg is sated," as science writer Natalie Angier notes. "It wants no more DNA."

Once the egg has been fertilized, the situation recreates itself—again and again, with spiralling complexity. Chemical messages instantly relay the news of the successful penetration to the brain. Molecules talk to molecules, cells to cells, organs to  organs. "Feedback loops" trigger a rush of specialized hormones, slippery secretions, and subtle muscle contractions to help ferry the fertilized egg to the uterus. Unlike Brewster's starfish eggs, the oocyte, as it's called, is itself in no hurry. For the first 3 or 4 days it divides, or cleaves, roughly once a day, slowing gathering into a tightly compacted ball. The ball is hollow. Shape being destiny, this tiny sphere—Brewster's "heap of soap bubbles"—anticipates perfectly the complex subdivisions to come.

Within a week the thicker end of the sphere glues itself to the uterine wall. Does the uterus "know" what's arrived? Clearly. The human immune system is programmed to distinguish between molecules that are "self" and "nonself" and to destroy the latter. The blastocyst, measuring less than one-hundredth of an inch across—a barely visible dot—behaves like a parasite, burrowing into the lining. Yet the uterus, after initially swelling to engulf the embryo and marshalling white blood cells to dispose of it, suddenly turns receptive, even acquiescent. Its blood vessels engorge with food and oxygen-bearing blood, and its tissues cordon off an area for the invader. Then, as the embryo bores through the small maternal blood vessels in its path, rupturing them, the hemorrhaging uterine tissues respond by releasing a starch that becomes its first meal. At once the embryo gorges itself and starts to grow at astonishing speed—doubling daily in size. Before the expectant mother knows she is pregnant, the basic relationship between mother and child is forged.

To understand what happens next—the budding of the embryo—recall Brewster's century-old descriptions. How within this tiny nestled dot does an embryonic human being organize itself and take form? How do "furrows" develop into spinal cords, "rods" into backbone? How does a descendant of an embryonic stem cell "know" to become part of an eyelash? Put another way, how does a half-brick know to become a steel girder or a credenza, then get itself to the 150th floor?

The scientific answer begins with a process called *gastrulation,* which occurs in all animals. Suddenly, a few days after implantation, the balled-up cells begin  to rearrange and move—migrate. With magnificent speed and coordination, sheets of cells stream past each other, some migrating inward, others out; some up, some down—until 3 layers emerge: inner, middle, and outer. From these relative positions, cells begin to change, or differentiate, into the building blocks for specific tissues, organs, and systems. The outer layer produces the cells for the skin and nervous system; the inner layer produces the lining of the gut and related organs like the liver. The middle layer starts to churn out cells that will become the heart, kidneys, gonads, bones, muscles, blood, and the rest of the viscera.

Once these layers are established, the cells interact rapidly with each other, rearranging themselves gradually into the complex communities we call organs. Poetic-minded embryologists compare this process to origami—the Japanese art of paperfolding. The instructions for rearranging a piece of paper involve only a few simple operations: folding and unfolding. No one would dream of making a crane's beak or a Minotaur's horns in one step from a single flat sheet. But by folding and pinching and refolding, then refolding again and again, it's possible to make ever more refined and complex shapes. Lewis Wolpert has been quoted as saying that it is not birth, marriage, or death, but gastrulation that is truly "the most important event in your life." He has a point. Everything we become, our whole magnificent design, is predicated on this early layering of cells.

It takes just 3 to 6 weeks to "lay down" the basic body plan. Still less than one-tenth of an inch from top to rump, the round ball of cells curls into a C-shape, a tiny comma. Already, as Brewster observed, a sort of groove has been cut along the length of the back, and it soon closes into a tube. Throughout the embryo, long stringy cells bundle to form nerves, which connect through the tube to the bulge at the top of the comma—the beginning 3-part brain. Two deep narrow slits form at the head—primitive eyes. To supply the rapid proliferation of cells with energy, a crude circulatory system unfolds. Again, the great triumphs are in shape and structure, just as with molecules. Another tube of new, highly elastic cells takes shape in the region below the dot. The cells have a remarkable ability to contract and release. Within 3 weeks from conception, they have twisted into an S-shaped loop and have started pumping cells filled with oxygen and nutrients to every other tissue in the body. A human heart begins to beat.

Furious folding and refolding happens everywhere, apace. At 6 to 7 weeks, small bulges on the embryo's flanks morph into recognizable limb buds. With neurons proliferating at the astounding rate of 100,000 per hour, the head grows swiftly. Eye pits deepen, the nasal region expands, the upper limb forms. Where a month earlier there existed a heap of bubbly cells now emerges a discernible face, no different from a chick's but a face nonetheless. The embryo weighs about the same as a raisin, yet there is little doubt where, biologically speaking, it is headed. Soon finger "rays" appear, and nipples.

And then, at about 8 weeks, just as quickly as it all began, the furious shuffling of cells stops. Assuming that everything has gone successfully, and most often it has, the not yet conscious embryo resembles a mature human being. The head, more rounded, measures about half the total body length, which itself measures about an inch and a half and weighs less than half an ounce. Still, the eyelids and outer ear are fully articulated. The tail disappears, eliminating the last primordial resemblance to other species. The liver, kidneys, lungs, and digestive system are all recognizable. In form, if not character, the embryo is complete, and to make the distinction goes by another name: fetus.

Much, of course, remains to be done. The fetus has to fill out, and its physical proportions must change to prepare for life outside the mother. But in a sense, the architecture is complete. By knowing "when and where to grow fast, and when and where to grow slowly, and when and where not to grow at all," the building has built itself.

But how has it done it? For Brewster, and for today's biologists like James Thomson, that remains the key question, the sum of them all.

We live in a remarkable time. Genetics and fertility research have taught us in recent decades how to conceive a baby. Now developmental biology and embryology are teaching us how to build one. Add to this the recent progress in seeing inside the human body — to zoom through the hidden structures of  organs, tissues, cells, even molecules, as if we ourselves were atom-sized — and we become witnesses in places we could scarcely imagine before.

You may wonder how this is possible: How can we look *inside* objects themselves too small to be seen? Most of the images in this book are not photographs, since photography cannot penetrate the surfaces of things. Instead, they are visualizations, composed through a marriage of powerful medical imaging techniques and passionate art. The core technologies are familiar — scanning devices like CT and MRI, and computer photoshop software. Yet combined, they illuminate a hidden world. A pea-sized embryo, for example, is scanned head to toe and the information compiled. Since the data reveal differences in density (cartilage is denser than liver tissue, which is denser than blood), it's possible, with the right software, to distinguish one organ, cell, and even atom from another. Using other digital techniques, the skilled artist can then isolate any object, magnify it, peel away its surface ("ramp down the opacity"), rotate it so it can be viewed from any angle, and add shading, shadow, and "pseudo-color" until its essence is revealed. In 3-D.

What's happening with the baby now? From what artist/scientists back to Leonardo da Vinci have suggested about the human form, and from what we can now *see* thanks to these penetrating images, we might better ask: *What's not?*

Conception

REPRODUCTION

GENES

FERTILIZATION

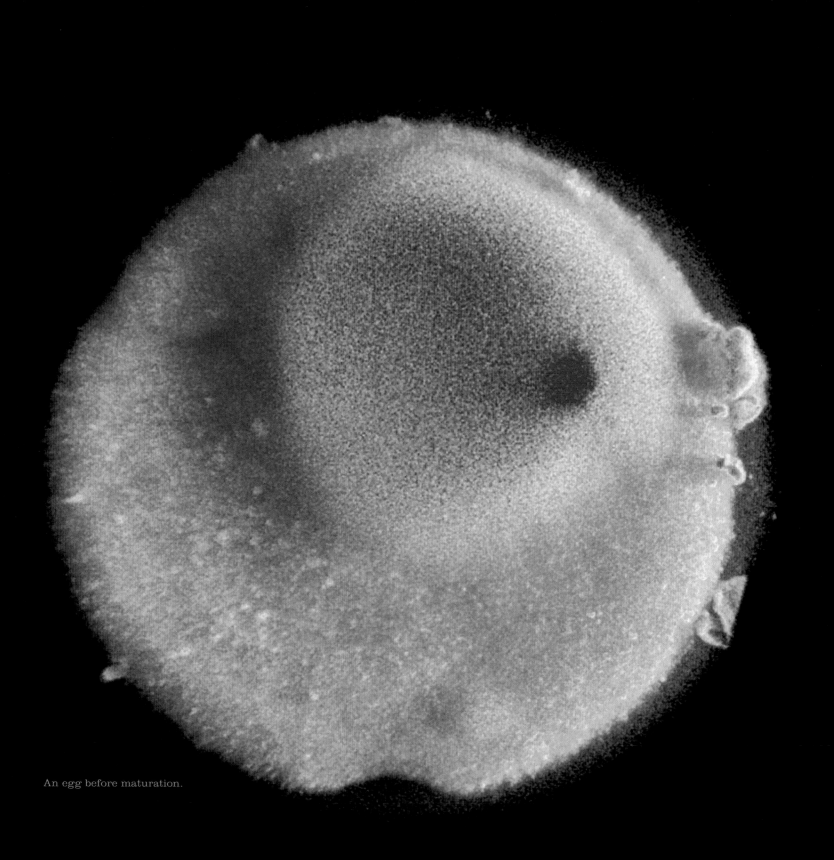

An egg before maturation.

01

Conception

A child springs forth, literally, out of what D. H. Lawrence called "pure blood consciousness . . . the consciousness of the night" — sex.

No wonder the ancients attributed the power of genesis to gods.

People have always tried to explain the origins of human life in special terms, terms that reflect what they understand about the rest of the world. The ancient Celts who prayed to storks, which they believed gathered up the souls of children from secret swamps and feeding grounds and delivered them to fortunate households, rationalized unwanted pregnancies by saying that the birds had eluded them, sneaking in at night and leaving them a child.

Among such creation stories, the hysteria following the discovery of the male sex cell in 1677 bears special note. The microscope had just been invented, giving scientists their first look at things too small to be seen with the human eye, and what some of them claimed to witness invoked both awe and rage. The father of the field, Antonie van Leeuwenhoek, said he saw in the semen of a patient suffering from venereal disease "animalcules," tiny men and women bent knee to brow and encased in translucent capsules — like "a small earthnut with a long tail," he wrote. "A million of them would not equal in size a large grain of sand."

The immediate response was extreme. Some learned scientists were bewildered that so many sperm would participate in the formation of a single human being, others were outraged by the pointless death of so many of God's creatures if, as was being posited, only one was responsible for fertilizing the woman. Eventually, the belief took hold that the animalcule was the "seed" and the female the soil into which new life was implanted. The animalcule's characteristics, including the sex, determined the baby's. This view remained scientific dogma for 150 years, until it was discovered that female sex cells also existed.

What wasn't understood was how these two cells, when joined, reproduced a new organism — a child. Scientists believed some animating force had to be involved, but the precise mechanism remained a mystery. Then a Moravian monk consumed by the new European obsession for cultivating flowers began doing breeding experiments on plants. Crossing different pea strains, Gregor Mendel showed that an offspring's characteristics reflected traits from both parents. Studying successive generations, Mendel proved further that there were rules governing how individual traits were handed down. Eventually he was able to predict a range of characteristics with mathematical certainty.

Mendel had discovered the science of heredity, and his findings quickly merged with those of the English naturalist Charles Darwin, who theorized that it wasn't only individuals, but whole species, that evolved through selective breeding. No one mentioned sexual reproduction per se, and working with human embryos was unthinkable, but there now emerged — toward the end of the nineteenth century — a modern scientific understanding of how human beings reproduce.

But what was the mechanism? How did it work? These were the questions that preoccupied early twentieth-century scientists as they set out to

A sperm can survive for up to 48 hours. It takes about 10 hours to navigate the female reproductive track, moving up the vaginal canal, through the cervix, and into the fallopian tube where fertilization begins.

solve the mysteries of world — not just biologists, but physicists and mathematicians who now revolutionized all science by enquiring into the behavior of the atom. In 1943, an exiled Austrian theoretical physicist living in Dublin, Erwin Schrodinger, began lecturing and writing on the subject, "What is Life?" Schrodinger deduced the crucial idea that genetic information must be stored at the molecular level — in other words, that heredity was carried out by small groups of atoms. Suspicion soon centered on the nucleic acids, those complex molecules packed within the nucleus of every cell. Since it was also now understood that what a molecule did was enabled by its structure, scientists went in search of an architecture that would explain how genes work. If it wasn't little men curled up inside spermatozoa, what was it? What did the material carriers of life look like?

"Too pretty not to be true," came the ardent verdict of one of the codiscoverers, James Watson. The story of how Watson, a 23-year-old former ornithology student, and Francis Crick, an ex-physicist, solved the structure of DNA — discovering that it was like a spiral staircase that could unzip down the middle to make copies of itself — has itself become mythic — Prometheus stealing fire from Olympus. But what it really did was to settle the question of where we come from. We come from our genes. As Watson himself boldly announced at the Eagle Pub in Cambridge, England, on a winter afternoon in 1953, he and Crick and their unacknowledged collaborator Rosalind Franklin had finally "found the secret of life."

DNA strands coil up, forming
spiral staircases. Each rung is a
micropiece of heredity.

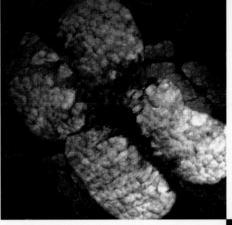

HEREDITY

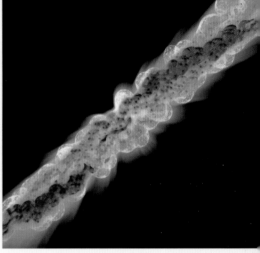

WHATEVER OUR DIFFERENCES as people, we all have in common a single fact: Our ancestors were all survivors. Generation after generation, our forebears have endured.without exception at least until early adulthood — until they successfully mated and had children of their own. From a hereditary standpoint, this is wonderful news. As we approach pregnancy and wonder what our own children will be like, and, especially, whether they'll be healthy, we can be reassured that their chances for survival are generally excellent, based on the legacy of successfulness bestowed on them by their ancestors

This is not to say that we are simply a product of our genes, but that we start life with the advantage of strong, select stock — that the genes we inherit are, to a remarkable degree, those that have been handed down by people who have been, in their own generations, among the best outfitted to survive and reproduce. As the sociobiologist Richard Dawkins puts it: "They have what it takes to become ancestors."

Now that it is our turn to become ancestors, we may be anxious about what we will hand down. Will our children be like us — or our parents, or grandparents — not only in the ways that make us strong, capable, and proud but in those that don't? Too, much can be made of our genetic predispositions; we are far too complex — too much the sum of lifetimes of living and learning — to be programmed by our DNA. Inheritance is not a zero-sum game.

What our chromosomes do carry, however, are building instructions, not for us but for the magnificent agglomeration of structures that enable us to become us. As the author E. B. White wrote: "Heredity is a strong factor, even in architecture. Necessity first mothered invention. Now invention has little ones of her own and they look just like Grandma."

The chromosome: It is a coiled multithread structure that carries the genetic code. A healthy body cell has 46 chromosomes arranged in 23 pairs.

Collagen protein: Proteins are necessary for growth and carrying out vital chemical functions. Collagen is the most common protein and the one we see in the mirror in the form of hair, skin, and eyes.

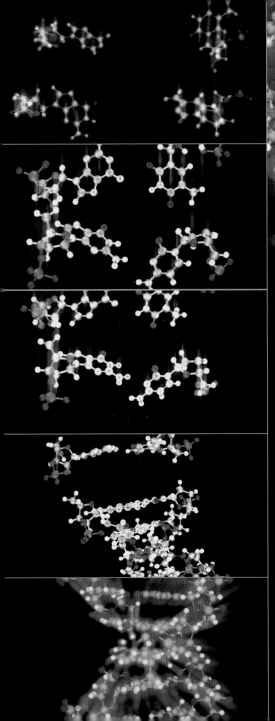

A section of DNA

Blueprint for a New Life
DNA

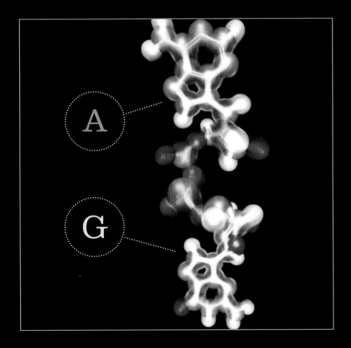

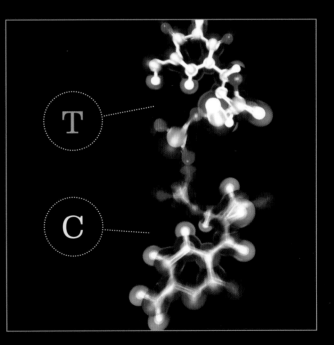

HEREDITY

The sequence on the left-hand page separates out a rung
of heredity containing the 4 nucleotides that make up
all of life. At our fundamental level these molecular struc-

INSTRUCTIONS FOR LIFE

Our DNA is made out of 4 components called nucleotides;
adenine (A), cytosine (C), guanine (G), and thiamine (T).

FATHER

AS CANDIDATES FOR ENVY, the male reproductive organs would seem to be wanting. Take the penis — "a ridiculous petitioner," the novelist William Gass has called it. "It is so unreliable, though everything depends on it — the world is balanced on it like a ball on a seal's nose. It is so easily teased, insulted, betrayed, abandoned; yet it must pretend to be invulnerable, a weapon which confers magical powers upon its possessor; consequently this muscleless inchworm must try to swagger through temples and pull apart thighs like the hairiest Samson, the mightiest ram."

And yet the grand plan for human reproduction could not want a more well-adapted anatomy — just as it couldn't wish for a more perfect, complementary reproductive apparatus in women. Everything about our sex organs is optimized to ensure the successful unity of sperm and egg.

In men, the concerted energies of several glands and organs combine to guarantee the successful transfer of healthy DNA during sex. The sperm are manufactured in the testicles, 2 egg-shaped organs that hang below the penis in the scrotum. Positioned outside the body wall because normal body temperature is too warm for sperm production, the testicles hang unevenly (so as not to jostle each other during ordinary movement) and fluctuate (a muscular response in the scrotum acts like a thermostat, winching them up and down to regulate their temperature). During the approximately 10 days it takes for them to mature, the sperm navigate coiled tubes in the epididymis, "storage tanks" that lie atop each testicle.

After that, the sperm are maintained in suspended animation in the vas deferens, a long tube connecting the testicles to a spigot behind the bladder. Remarkably, they sedate themselves during this period to conserve vitality — they release carbon dioxide, creating a mildly acidic environment that temporarily paralyzes them. Then, just before ejaculation, several glands (the prostate, Cowper's glands, and the seminal vesicles) release substances to fortify the sperm for their journey and lower the natural acidity of the vagina. So highly prized is their mission that the sperm are packed with nature's sweetest sugar, fructose, which the body manufactures only for this purpose.

And the penis? Well, perhaps it's not so ill-suited after all. Among mammals, only primates have penises that hang freely outside the body wall — so that they can urinate efficiently from trees, a clear evolutionary advantage over other species. And among primates, only man doesn't have a penile bone, which could be easily fractured and incapacitated as he stands upright and walks on two legs. The columns of spongy tissue that swell with blood during erection offer not only utility but pleasure, encouraging repeated use. In other words, though it may behave absurdly and look silly, the penis — like heredity itself — has evolved to help maximize successful reproduction.

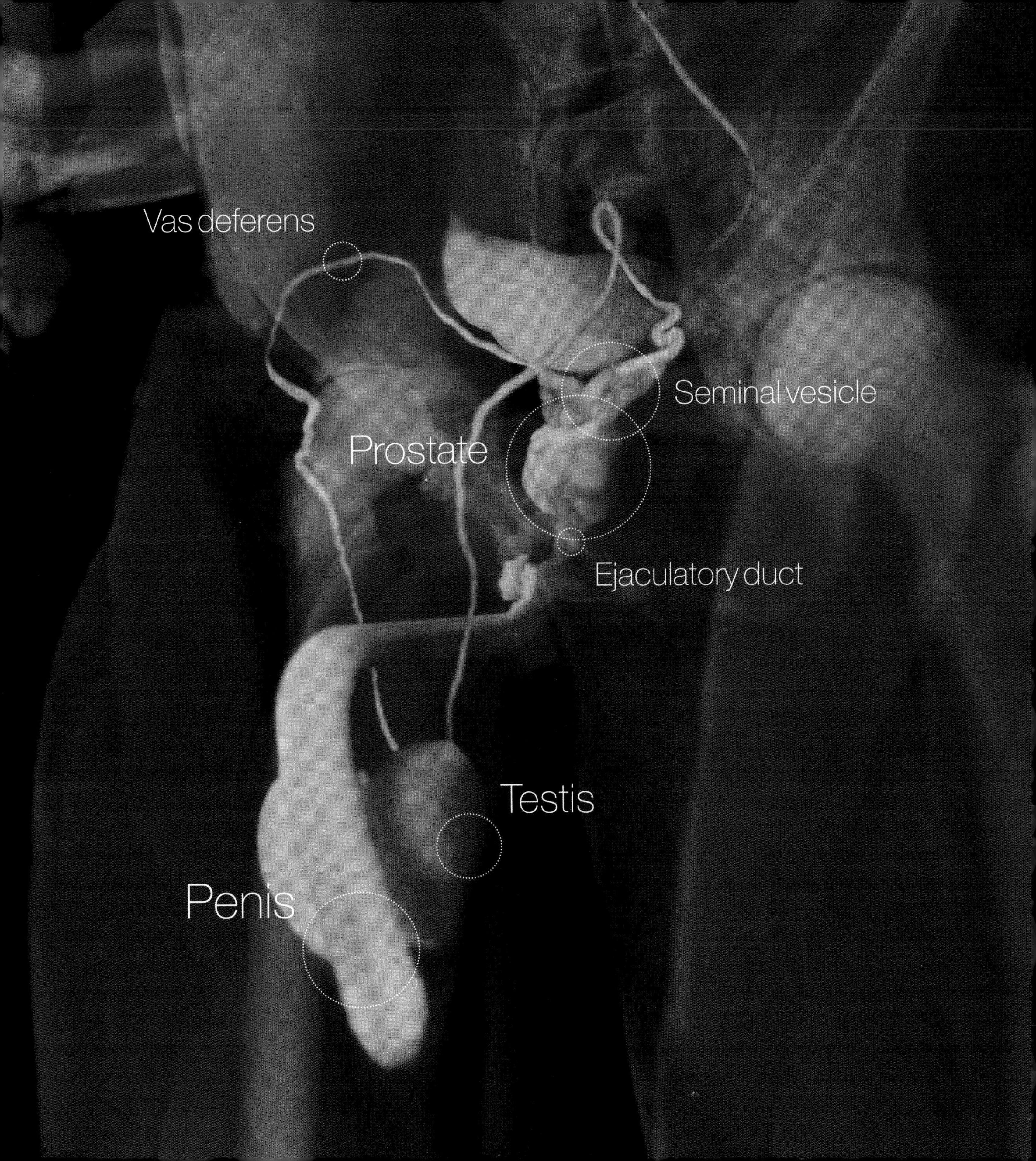
Vas deferens
Seminal vesicle
Prostate
Ejaculatory duct
Testis
Penis

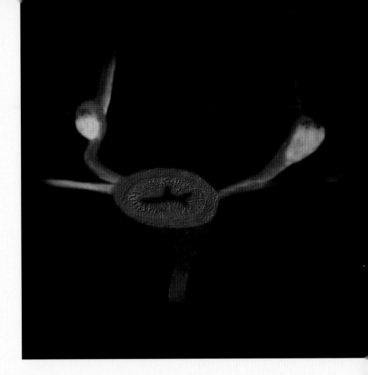

MOTHER

BUILT OF SKIN, MUSCLE, AND FIBROUS TISSUE, the vagina is a "pause between the declarative sentence of the outside world and the mutterings of the viscera," the author Natalie Angier writes. It obliges in both directions, a cannily engineered marvel that helps deliver sperm to egg, then 9 months later accommodates the delivery of a child. Her role in reproduction more complex and more sustained than a man's, a woman's reproductive system needs to be not just a collaboration between organs but a sustaining environment, an ecosystem. The vagina is its elegant vestibule. "A Rorschach with legs," Angier notes of its special ambiguity. "You can make of it practically anything you want, or need, or dread."

As a sex organ, the vagina is more than a simple void, a 4-to-5-inch-long tunnel extending from the outer opening, the labia, to the cervix, the donut-shaped ring of muscles that guards the uterus. Its anatomy makes up the body's most intriguing bit of fleshy origami. From the mound of fatty tissue that cushions the pelvic bone during intercourse, the mons pubis, it parts below to reveal the genitals. Two liplike concentric folds of tissues, the labia major and labia minora, protrude from the opening;

Each month an ovum is released and travels down the fallopian tubes to the uterus. If the egg is not fertilized, menstruation occurs.

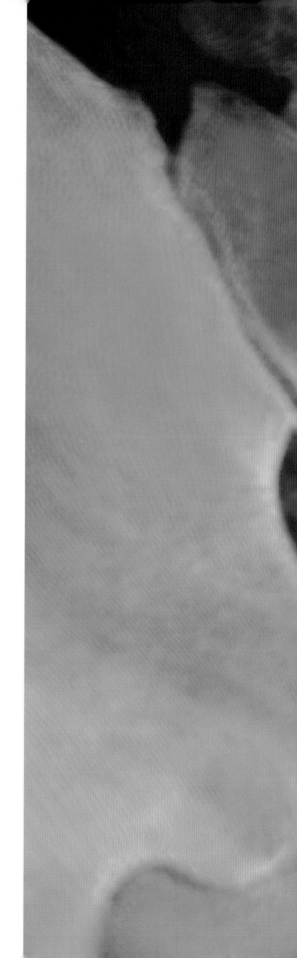

the first protects, the second lubricates. Buried in the muscle and fibrous tissue at the vortex of the inner lips is a veiled bud of blood- and nerve-rich tissue — the clitoris.

It is this last that has lately sparked the most interest and controversy, with the most fervent debates having to do with whether or not it is biologically essential. Women, unlike men, can reproduce without orgasms. Yet the pea-sized clitoris, a center of sexual pleasure, contains a charged hive of 8,000 nerve endings — twice as many as the much larger penis — the most of any external structure in the body including fingers, lips, and tongue. Plainly, it is meant to feel things, and deeply.

Beyond the cervix lays the core reproductive apparatus, the uterus and ovaries. Two almond-shaped organs similar in size and shape to the testicles, the ovaries produce female hormones, which induce physiological changes like breast growth. They also house and nourish the ova, the eggs, which, once ripened, are released into the fallopian tubes, where they may encounter sperm. Once an egg is fertilized, it is designed to continue its migration into the hollow, muscular, pear-shaped uterus. It is within the containment of the uterus that the fertilized egg takes up life, developing until it is forcibly expelled, some 38 weeks later, through the vagina and into the waiting world.

Right: After fertilization the egg travels down to the uterus, which provides protection and livelihood for the baby during the 9 months by great expansion.

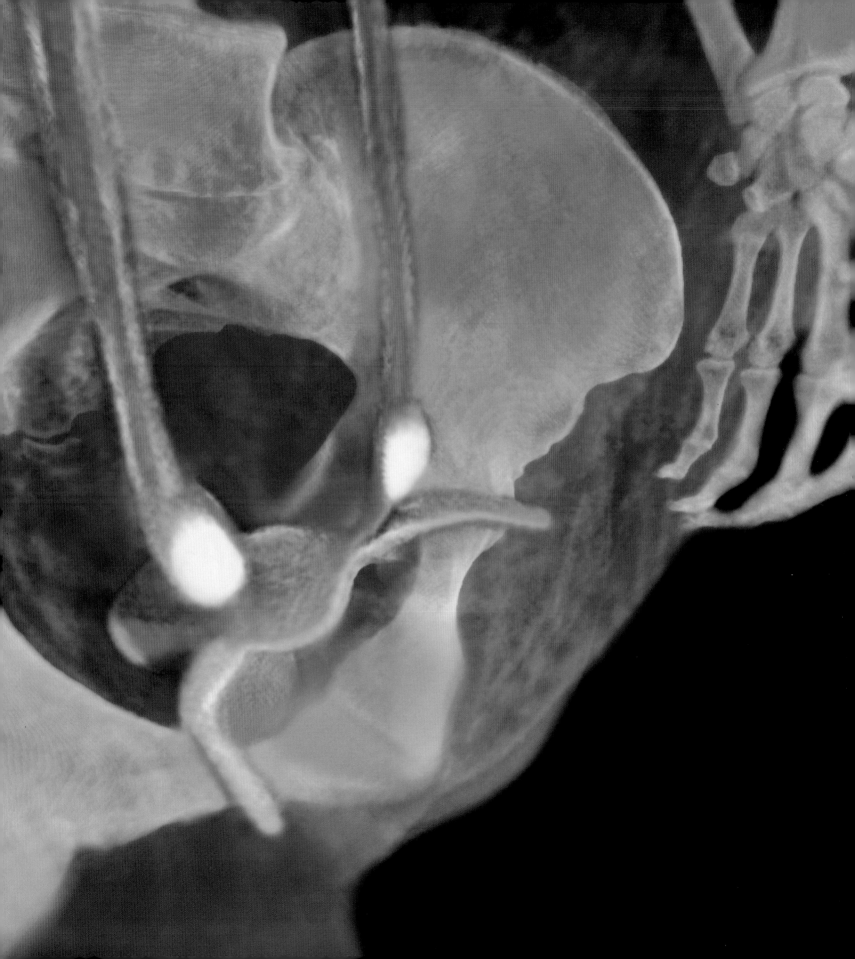

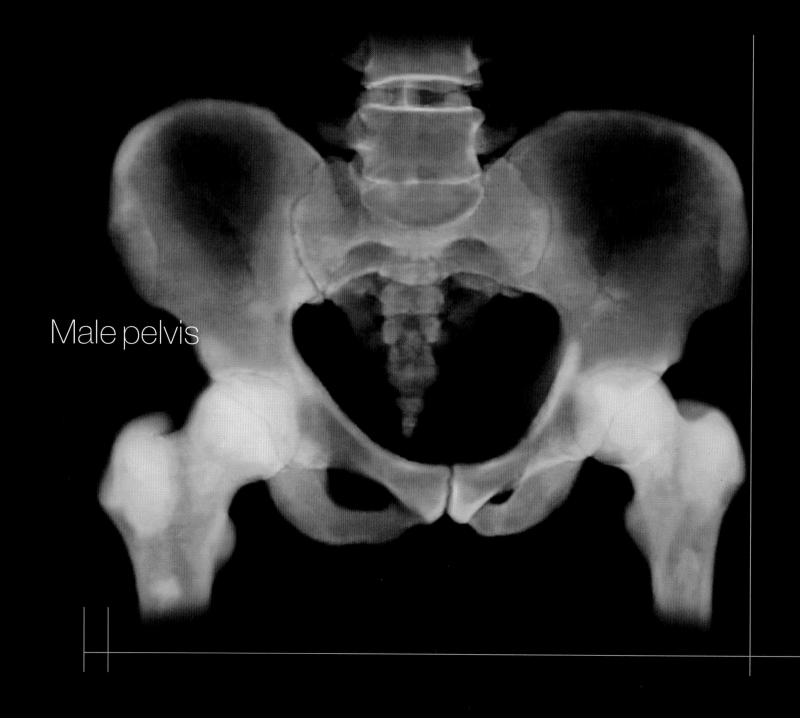

Male pelvis

Designed to Carry a New Life

Shallower pelvis

Female pelvis

The encircling shape within the pelvic bones is wider and shallower in the woman, allowing the uterus to grow. Another noticeable difference is the much wider angle at the bottom of the pelvic bones.

Wider pelvic arch

ANATOMY OF THE EGG

OF ALL THE FEATURES built into the human reproductive cycle to safeguard the future of the species, perhaps most critical is the length of our fertile periods. Women can bear children for 30 years or more, men often for as long as they live. It has been observed that if every opportunity for pregnancy was seized and if a woman somehow retained her health and sanity, she could become a mother more than 30 times over.

The key limiting step is the maturation once a month of a single ovum. Of the 6 to 7 million eggs first produced in the female during gestation, all but 40,000 are destroyed naturally before puberty. From then on, the future is laid down through the serial selection, care, nourishment, and discharge of just some 400 ova.

This process—ovulation—starts with a wave of hormone released from the pituitary, the eye-shaped gland at the base of the brain that directs the endocrine system. Inside the ovaries, each egg cell nestles in its own pod, or follicle. Spurred by the wash of stimulating hormone in the blood, between 15 and 20 follicles begin secreting estrogen, the female sex hormone, which, among other functions, awakens and nourishes undeveloped eggs. At day 10 or so, for reasons not yet understood, a single egg is selected to ripen fully.

Four days later, another stimulating hormone is released from the brain, the active follicles split open, and the eggs sail out into the fallopian tubes, which suck them in. (Sometimes blood flows from the tiny rupturings, causing spotting and cramps.) The contents of the 19 or so incompletely ripened sister follicles shrink and die as the one survivor brushes along the wall of the tube, awaiting fertilization.

Of all the body's cells, the mature ovum alone is spherical. "The form makes sense," Angier observes. "A sphere is one of the most stable shapes in nature. If you want to protect your most sacred heirlooms—your genes—bury them in spherical treasure chests." More, it is surrounded by a halo of helper cells—the *corona radiata*—which nurse and protect it, and a thick shell-like matrix of sugar and protein, the zona pellucida, which acts as a kind of bouncer, restricting entry.

Perhaps 40 or more years since it was first produced, the female reproductive cell is ready for her close-up.

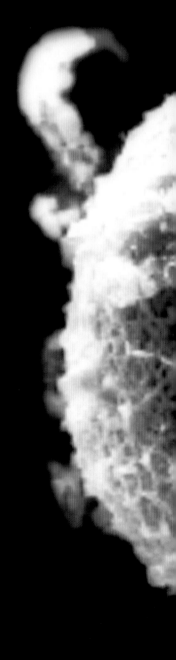

The egg, at a much higher magnification. Note how dense and thatched the outer membrane is, which the sperm must penetrate.

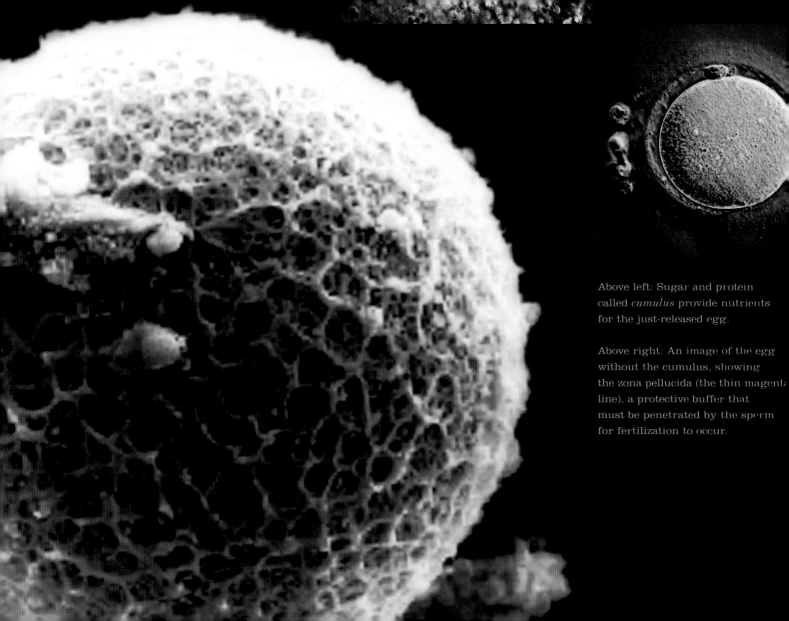

Above left: Sugar and protein called *cumulus* provide nutrients for the just-released egg.

Above right: An image of the egg without the cumulus, showing the zona pellucida (the thin magenta line), a protective buffer that must be penetrated by the sperm for fertilization to occur.

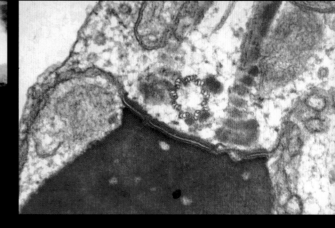

Left: Cross section of the head of
a sperm containing 23 chromosomes (the father's contribution to
the new life).

Above: Structural view of a cross
section of the base of a sperm's
head.

Facing page: Cross sections of
sperm tails. Inside the tail, small
but strong coil-like structures provide the motion to swim.

ANATOMY OF THE SPERM

LOTTERY JACKPOTS, AT 80 MILLION TO ONE, have better odds. Every second, the typical adult male produces 1,000 sperm. Over a lifetime, he may produce as many as 12 trillion sperm. There are about 300 million sperm in a cubic centimeter of semen, the average ejaculate.

Why so many? The answer: Because only such an extravagant surplus can reasonably ensure that one of them will reach their shared goal.

The function of the male reproductive cell is to make contact with the ova in the fallopian tube, penetrate it, and deliver its tiny packet of healthy DNA. Yet the extraordinary difficulty of the endeavor guarantees that only a handful will succeed.

Design-wise, the sperm is an uncommonly sleek vehicle. Its ovoid head wears a protective cap containing an enzyme to digest the cumulus cells around the egg. Its braided midsection attaches head and tail and contains the power pack. Its familiar whiplike tail tapers to a tassled filament, for added locomotion.

More than most cells, however, the real zone of interest is the nucleus. In humans, chromosomes come in pairs of 23. Yet in anticipation of their combining, sperm and egg have only half that number. In sperm, uniquely, all of the chromosomes either are X-shaped or Y-shaped. These variations determine a child's sex: X for girls, Y for boys. Because X and Y sperm are manufactured in equal number, we might expect equal numbers of male and female embryos, but statistics reveal that 106 boys are born for every 100 girls. Scientists speculate that Y sperm swim faster than X sperm because they carry slightly less genetic material, accounting for the disparity. As for the larger evolutionary purpose of producing more boys than girls, some biologists suggest it is to compensate for higher mortality rates among young males. By the time the reproductive years begin, the male to female ratio is even.

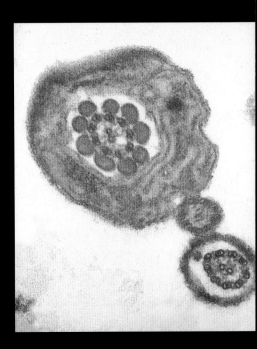

COPULATION AND ORGASM

— Nora Ephron, *When Harry Met Sally*.

Claudius Galen, the most celebrated physician of ancient Greece, believed that women are just men inside out and that both sexes needed to have orgasms to conceive. Though he was wrong on the second point, Galen understood correctly that the union of man and women is based on mutual fulfillment.

Why two people desire each other sexually is a question perhaps best answered by themselves and by poets. Strict evolutionists suggest that in choosing a mate we are selecting a DNA profile for our (desired) children. What seems likelier, if conception is the goal, is that the *idea* of having a child has eclipsed all else, and what a couple hopes for is literally to merge, by sowing a new life.

The physiology of human mating starts with arousal. As we become excited, our sex organs prepare for coitus through a series of changes in the blood and nervous systems. The brain, receiving signals from the genitals, is tantalized, anticipating erotic pleasure. Our hearts beat harder, flooding our arteries, while our veins constrict. Blood engorges the erectile tissue of the penis and clitoris as well as the testicles, ovaries, and labia minora. Our muscles tense. Nipples stiffen.

In time, these effects plateau. In the woman, the outer third of the vagina becomes vasoconstricted, while the inner two-thirds expand slightly and the uterus becomes elevated—all in preparation for receiving sperm. Not our exertions but our involuntary nervous systems now increase our breathing and heart rates.

Orgasm occurs with the rhythmic involuntary pulsing of the sex organs, a shuddering rapture. In men, this occurs in 2 stages. As the intensity builds, reflex centers in the spinal cord send impulses to the genitals, prompting the smooth muscles of the testes, epidemymes, and vas deferentia to contract and squeeze sperm into the urethra. There, all traces of urine, which can kill sperm, have already been neutralized by glandular secretions. The prostate gland and seminal vesicles add their own secretions to the mix. With emission, the man loses control. The filling of the urethra triggers the muscles encasing the base of the penis to contract, and they force the semen out.

Women's orgasms involve the uterus, the outer vagina (including the clitoris), but not the upper two-thirds of the vagina. Typically harder to stimulate to orgasm than men, yet capable, once aroused, of multiple orgasms, women appreciate continued, gentle stimulation.

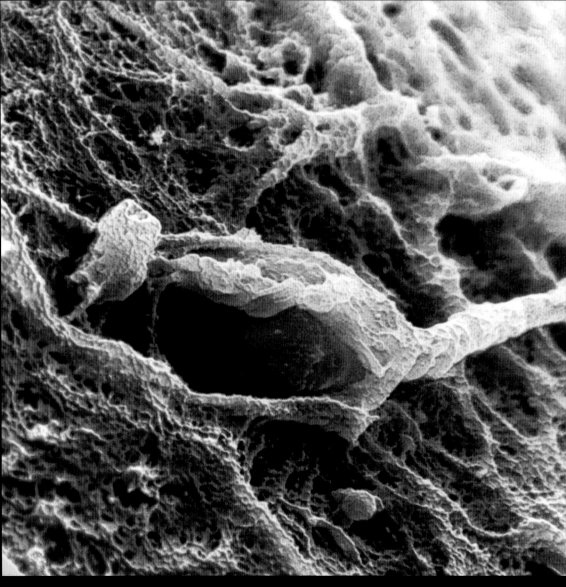

The first one getting through the wall of the ovum fertilizes the egg.

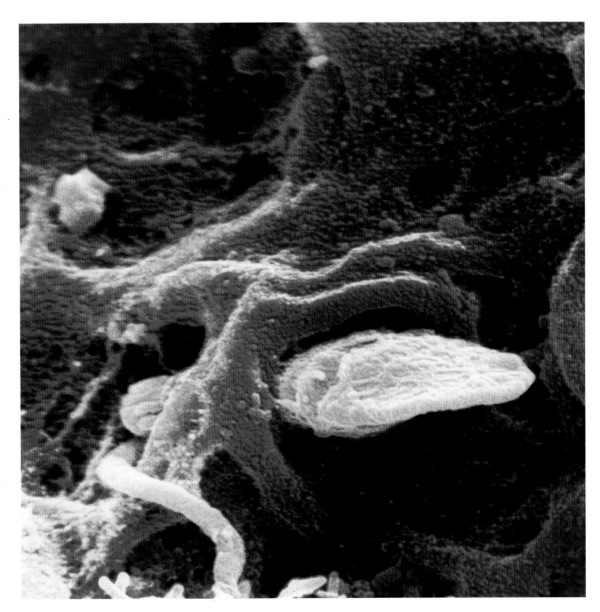

Survival skills rewarded. One in a few million.

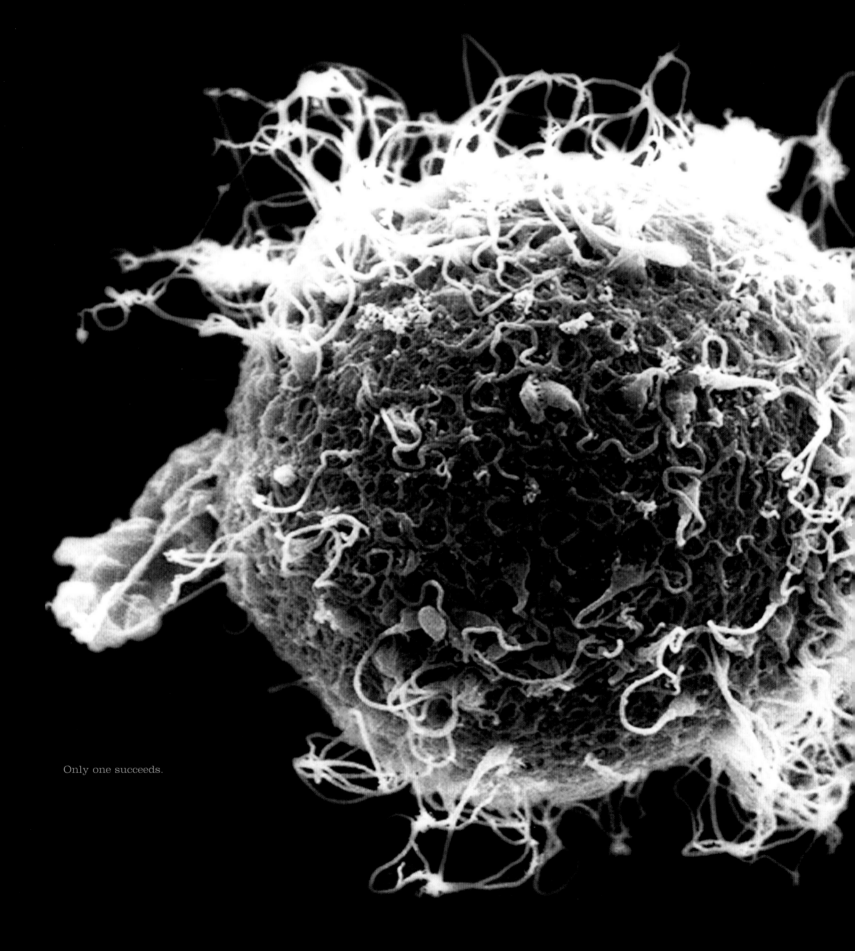

Only one succeeds.

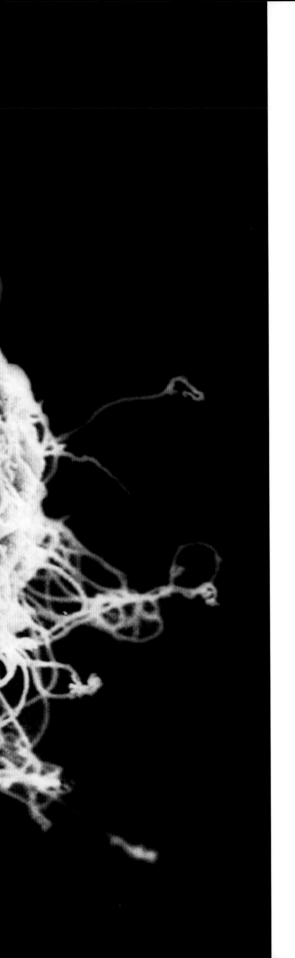

FERTILIZATION

THOUGH 300 MILLION SPERM may enter the upper part of the vagina, only 1 percent, 3 million, enter the uterus. Only one sperm needs to bind with the protein receptors in the zona pellucida to trigger an enzyme reaction, allowing the zona to be pierced and the sperm to get in and fertilize the egg. Penetration of the zona pellucida takes about 20 minutes. The fusion of sperm and egg is not a single event but a series of winnowings. Even if coitus occurs at a propitious time — not too long before or after ovulation — newly released sperm still face a succession of lethal pitfalls.

Think spawning salmon.

The first barrier — though hardly the most daunting — is distance. Whipping their tails, sperm swim mightily, about a half-inch per minute. But the upper reaches of the fallopian tubes are nearly a foot away — in fish terms, miles.

In between, the sperm must navigate a deadly gauntlet of physical and chemical obstacles. Acid kills sperm, and though the normally acidic vagina becomes less so during ovulation, most male sex cells would die there if not for the protection of the semen, which is alkaline. Still more resistant is the cervix. Although its sheathlike mucous plug thins at ovulation, the area remains thick with fibrous cells that block and ensnare millions of migrating sperm. Scientists compare it to the Sargasso Sea.

Those that survive and enter the uterus soon face a choice that for half will mean certain failure and extinction. Only one ovary has generated a mature egg, and if a sperm enters the wrong ovaduct, it's tantamount to a salmon struggling upstream in a barren tributary.

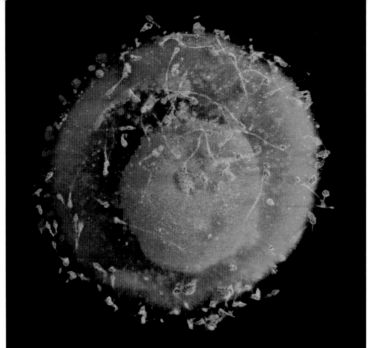

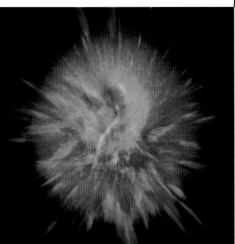

Fertilization: The process begins as the sperm penetrates an oocyte (an egg). The fertilized egg then becomes a zygote. The zygote then begins to descend into the uterus from the fallopian tube. Descent takes approximately 3 days.

➤ Facing page: The fertilization process takes about 24 hours. Rejected sperm will surround the egg for many days.

Strikingly, the harshest travails await those that enter the correct passageway. The waving cilia and muscle contractions that propel the egg downward beat back the current of sperm. The walls are richly variegated, and many sperm become entrapped. Scavenging white blood cells, ever alert to intruders, attack and destroy vulnerable sperm. Of the hundreds of millions of sperm released during ejaculation, fewer than 500 — perhaps one in 500,000 — finally encounter the ovulated egg. Of that number, barely a handful breach the outer shell and bind to the inner membrane.

And then, if success is to be obtained, comes the final selection. Once the first sperm penetrates and fuses with the egg, the membrane rapidly changes electrical charge, in effect demagnetizing. All other competing sperm literally drop off. Less than five minutes later, a second, more permanent chemical blocking mechanism is triggered, safeguarding the unique mixture of parental genes. In reproduction, nature abhors threesomes. Still, like jilted suitors, the excluded sperm continue to flutter around the ovum for several more days.

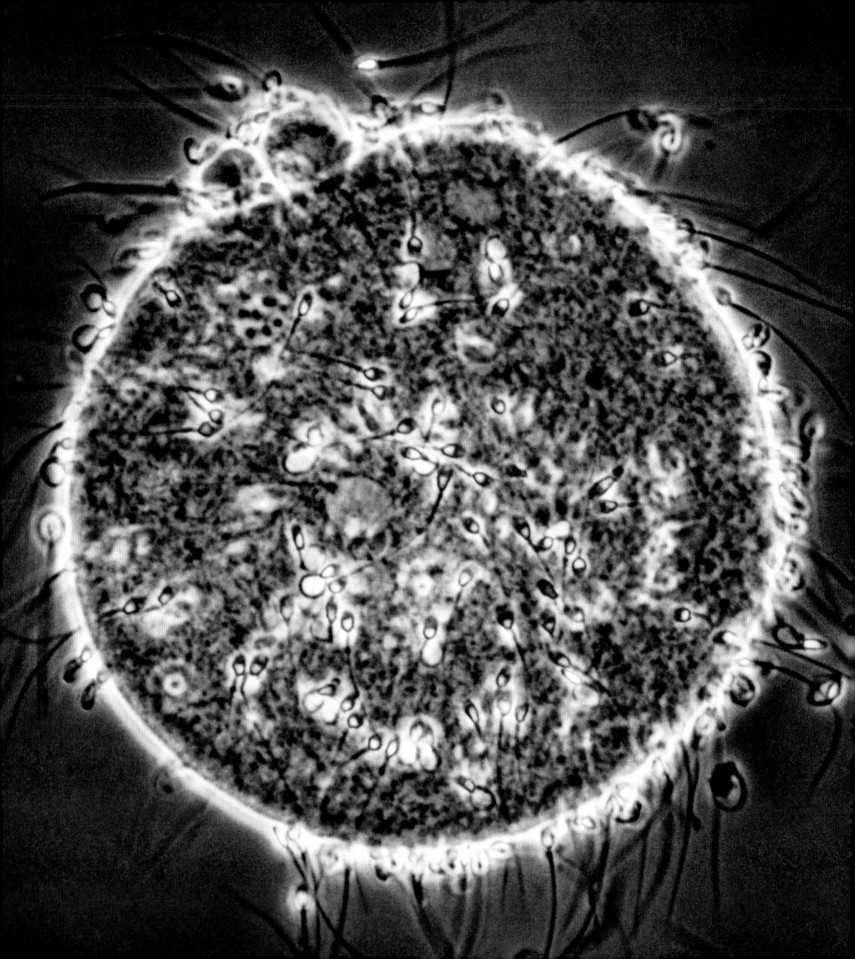

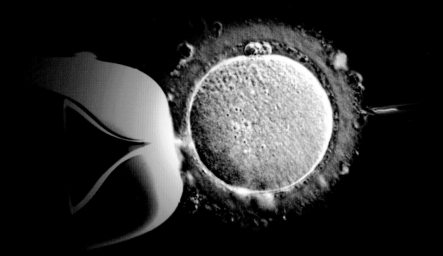

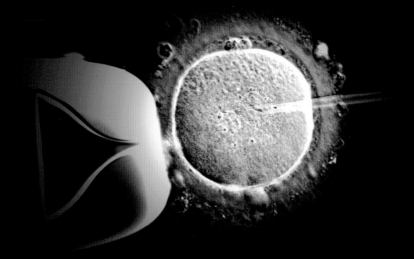

ARTIFICIAL INSEMINATION

One form of artificial insemination
involves injecting a single sperm
directly into the oocyte.
Top: Close-up of the needle
penetrating the oocyte.
Middle: The process involves
injecting just one sperm, replicating
natural fertilization.
Bottom: Note tail of sperm in circle.

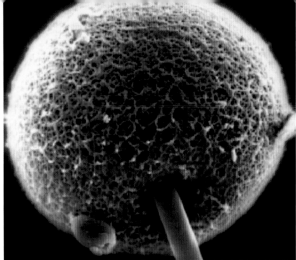

INFERTILITY

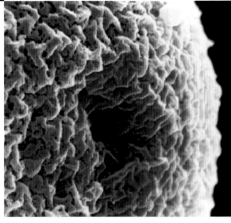

GIVEN THE OBSTACLES TO SUCCESS, it's not surprising that many
couples have trouble conceiving. In such a complicated process, much can—and
does—go wrong.

Many men do not produce sufficient sperm, or their cells are weak swimmers
or lack "purpose"—a sense of where to go. Others produce too many sperm cells that
are malformed or die before they reach the egg. Some men are infertile because of
genetic diseases like cystic fibrosis or chromosomal abnormalities.

Women commonly have problems ovulating, or their fallopian tubes are blocked,
or their eggs are unhealthy.

None of this is cause for shame or alarm. Infertility is a medical problem, not a
sexual disorder. In the U.S., more than 5 million people of childbearing age experience
infertility—about 10 percent of the reproductive age population—with men and women
affected equally. At least half these people will respond to treatment with a successful
pregnancy. Among couples seeking medical help, most—85 to 90 percent—receive
conventional therapies such as drugs or surgery, the rest high-tech aids like in vitro fertilization or the use of donor eggs.

Couples are generally advised to seek medical counseling if they can't achieve
pregnancy after a year of unprotected sex.

DIFFERENTIATION

With Child

THE FIRST CELL

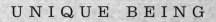

UNIQUE BEING

02

The First Month

Fifty years before Edwin Tenney Brewster described an embryonic architecture of rods and furrows in his "child's" book on life and sex, the nineteenth-century British social theorist John Ruskin offered his own view of how humans take form: "You may chisel a boy into shape, as you would a rock, or hammer him into it, if he be of a better kind, as you would a piece of bronze. But you cannot hammer a girl into anything. She grows as a flower does." Now, a century and a half later, biologists explore the question of how we are built by examining the stuff we are built of, our molecules. The wonder is not that boys and girls are made, in similar numbers, but that their smallest, most primitive parts know from conception how to make them.

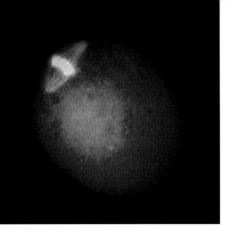

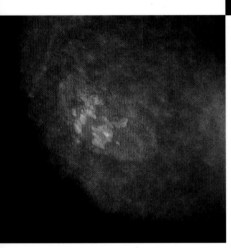

FIRST OF TRILLIONS

After duplication the chromosomes exchange DNA (they cross over) in a unique and random way that cannot be re-created.

A new mixture from the mother and father has been created. Now the cell has double the amount of chromosomes it needs. It has to divide into 2. Little threads called *spindles* pull each chromosome from the middle to the sides.

MITOSIS: Inside, one of the the most thrilling and important—and routine—of all biological events slowly beings to unfold: cell fusion. Through some force that is not understood, the egg's protoplasm starts to shimmy, violently. The nuclei of sperm and egg sidle towards each other, enlarge, and shed their protective membranes.

Within 12 hours, the nuclei merge, followed by the commonest of miracles. The 23 maternal chromosomes and 23 paternal chromosomes attach, creating the first edition of the 46-volume set of instructions for turning this one cell into the trillions that will make a complete individual. From now on, every time the daughter cells of this cell— now called a zygote—split in 2, each will carry a perfect copy of this blueprint.

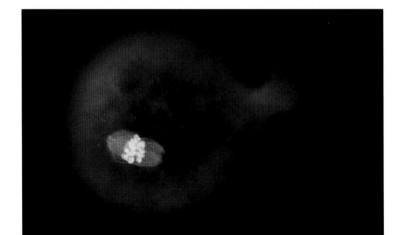

The genes from the parents get exchanged. Each sperm and egg carries 23 chromosomes (half of what a human has). The process starts with duplication of each member of the 23 chromosome pairs.

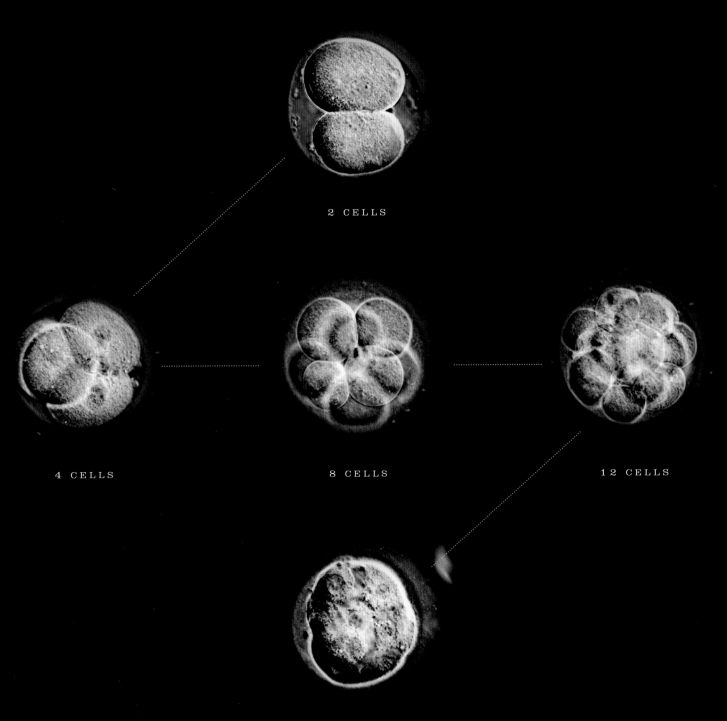

2 CELLS

4 CELLS

8 CELLS

12 CELLS

16 CELLS

MAKING COPIES

BEFORE A CELL CAN SPLIT IN TWO, its chromosomes must do the same. Along each chromosome, a thousand or more genes (each containing millions of atoms) are strung like pop beads on a chain. Within a few hours after the nuclei of sperm and egg have fused, every gene synthesizes a copy of itself from available chemicals in the zygote. Then each chromosome splits lengthwise, creating two half-chromosomes that regroup into 2 distinct nuclei.

To watch cells divide, or "cleave," is to witness a monumental improbability. One minute there is a spherical mass of protoplasm. The next, the ball pancakes on one side, surface tension gathers, a pinched line appears in the middle, and suddenly the mass parts as if pulled by a drawstring into two equal halves, each within its own membrane and with its own nucleus. Distinct structures, the cells stick together, although occasionally they will split apart and divide independently into separate pairs of daughter cells—identical twins. (Fraternal twins result when 2 separate ova, usually one from each fallopian tube, are fertilized.)

By 2 days after conception, the cells cleave again; by the end of the third day, after splitting once or twice more, 16 to 32 cells huddle inside the zona pellucida, like tiny soccer balls jammed in a clear sack. Each ball is identical, each subdivides at the same time as the others. Feathered along by millions of hairlike cells lining the walls of the fallopian tube, this clump—now called a *morula*, Italian for mulberry—floats towards the uterus. If the cluster becomes trapped in a fold, there is the risk of a tubal, or ectopic, pregnancy, although the great majority pass through safely.

◄ Facing page: After the chromosomes are pulled aside, the cell divides (cleaves) and becomes 2 cells. The zygote's first cell division begins a series of divisions, with each division occurring approximately every 20 hours. Spindles pull apart again as the 23 pairs of chromosomes and the cells divide. The result is now 4 cells that have a uniqe set of 23 chromosomes with a DNA that comes from both parents. When cell division reaches about 16 cells, the zygote becomes a mulberry-shaped morula.

MOMENTOUS CHANGES

4 DAYS

➤ Facing page: About 4 days after fertilization, the entire structure is now called a *blastocyst*.

SO FAR, THE CELLS THAT WILL BECOME YOUR BABY have behaved no differently from each other. If that were to continue to be true, the result would be an endlessly subdividing ball. But now several vital changes take place, and the sum is towering. Darwin identified the primary laws governing life as "Growth with Reproduction, Inheritance, and Variability." As the morula enters the open cavity of the uterus, where it will develop during pregnancy, all three laws arise with dramatic effect.

First the morula alters shape, its cells subtly shifting location. Instead of a dense cluster, they now form a single-layered hollow sphere around a fluid-filled cavity. The way an individual cell comes to have a specific purpose, or grow to a predictable size and shape, or move among its neighbors as if it knew what it was doing, is determined by which of its genes are switched on and which off. In this new formation, known as a *blastocyst,* cells begin to "differentiate" — act unalike.

Two types of cells start to form, with profound implications. Some of the cells around the hollow ball clump together on one side. These will become the child. The rest (the outer ring) will develop into the child's environment — the membranes that will protect, nourish, and contain it, and connect to its mother.

Hours later, a second major differentiation occurs — this time within the inner disclike cluster. The cells become layered, just two thick, yet the layers themselves represent another major division. The outer layer — the ectoderm — contains cells that are destined to develop into the brain, spinal cord, nervous system, sense organs, and skin; the inner layer — the endoderm — will become the inner lining of the gut, including the stomach and intestines. Later a third layer — the mesoderm — will emerge between them, from which will arise the muscles, skeleton, and most other internal organs.

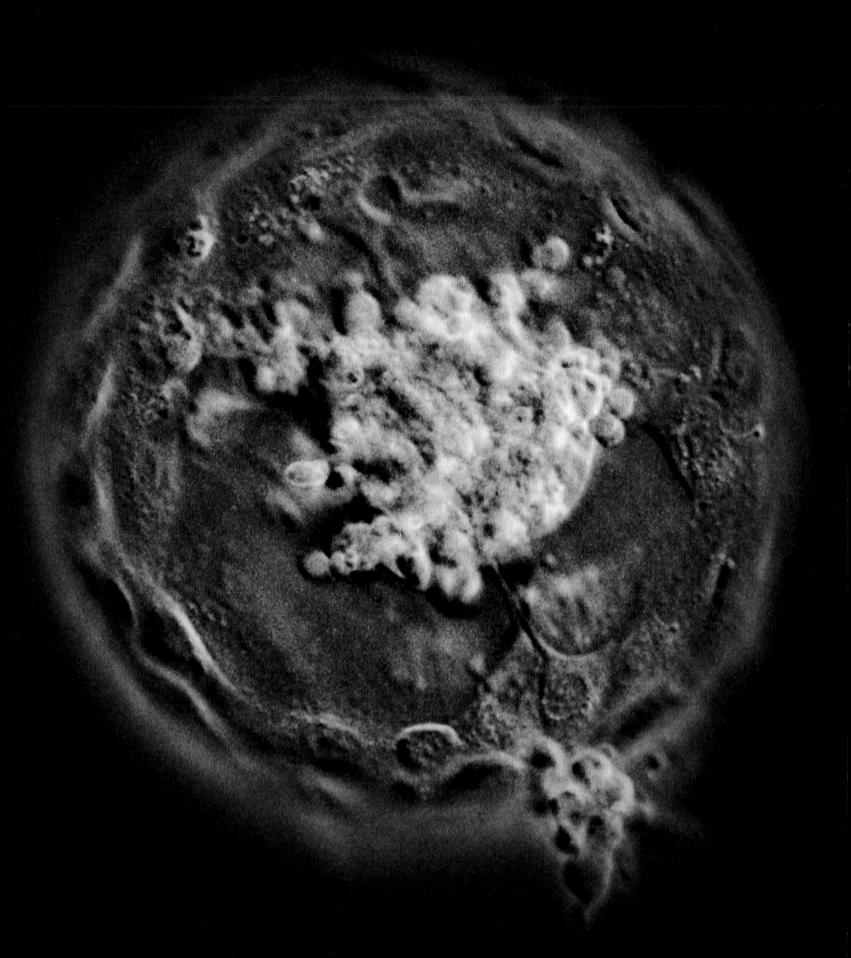

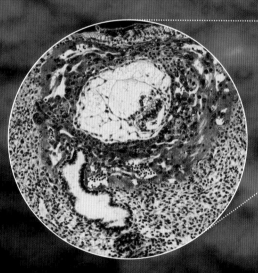

The embryo mass divides, rapidly
forming a two-layered disc. The top
layer of cells will become the embryo
and amniotic cavity, while the lower
cells will become the yolk sac.

WELCOMING A NEW LIFE

BARELY A FEW HUNDRED CELLS, the blastocyst measures about one-tenth of a millimeter — the width of a medium penpoint. Yet if it is to keep growing and changing, each cell needs an ongoing supply of two things: instructions (where to go, when to subdivide next) and, especially, nourishment.

Since fertilization, each cell has lived off stored energy and newly minted DNA; there's been no further help from parents, and the tasks have rapidly become much more complicated. Its metabolism (the sum of its hugely complex chemical activities) is doubling every few hours. To survive, the blastocyst must organize to feed all its cells and be fed by its mother.

Meanwhile, though a woman won't know yet that she's pregnant, her body also prepares to be transformed. Biologically speaking, she is about to become a host.

Tumbling through the uterine cavity, the blastocyst hatches from the zona pellucida, which allows it to expand and release the cells on its surface to interact with the outside. At the same time, hormones from the abandoned follicles in the mother's ovaries bathe the lining of the womb, making it soft, porous, and absorbent — receptive. When cells along the rim of the blastocyst, now secreting chemical enzymes designed to erode the already softened uterine lining, make contact, they latch on. The process is called implantation, with the ideal site located on the back wall of the uterus, near the support and protection of the spine.

In preparation for implanting, the blastocyst develops a basic system for passing molecules from cell to cell.

Left: 7-12 days after ovulation, cells of the uterine lining are destroyed, creating blood pools, foretelling the growth of the placenta.

SYMBIOSIS

THE JOINING OF 2 LIVES that now occurs is the second great challenge of pregnancy, after fertilization. For nearly a week, the construction of a system to nourish several hundred cells so that they can grow into an 8- or 9-pound baby takes priority over all else.

Unfolding in unison, the mechanics are a marvel. While machetelike enzymes on the rim cells of the blastula cut into capillaries in the uterus, rupturing them, the cells themselves fuse into a tough membrane called the *chorion,* which for 9 months must surround and protect the child. Once the blastocyst has burrowed deeply and is awash in the mother's blood — rich with starch released during hemorrhaging — fingerlike structures from inside the chorion reach out and tap the mother's circulation. (Parents might wonder if vampire myths originate in an unconscious recollection of this earliest mother-child encounter.)

The most important instrument aiding this growth and transformation is the placenta, a disc-shaped mass of tissues that forms between the chorion and the uterine wall and serves as an "exchange organ." Vessels from the mother and embryo intertwine without joining, allowing stale blood to be exchanged for fresh, waste products for nourishment, embryonic hormones for maternal hormones — while barring passage of infection and most other harmful agents. The emerging relationship between mother and child — mutual, nourishing, deeply interconnected — is forged in this crucible of shared cells.

Meanwhile, on the opposite side of the disc, a deposit opens and a new structure forms — the yolk sac. In birds, yolk is food for the egg, because the bird doesn't share its mother's circulatory system; it needs a nutrient supply that will sustain it all through gestation. The human yolk sac contains no real yolk but serves a similar function, producing primitive blood cells for the embryo until it can develop organs to do so: the liver, spleen, and bone marrow.

Amazingly, the yolk sac also gives rise to the sex cells. Barely a week old and consisting of less than 1,000 cells, the embryo is already preparing to replicate — 20 or more years into the future.

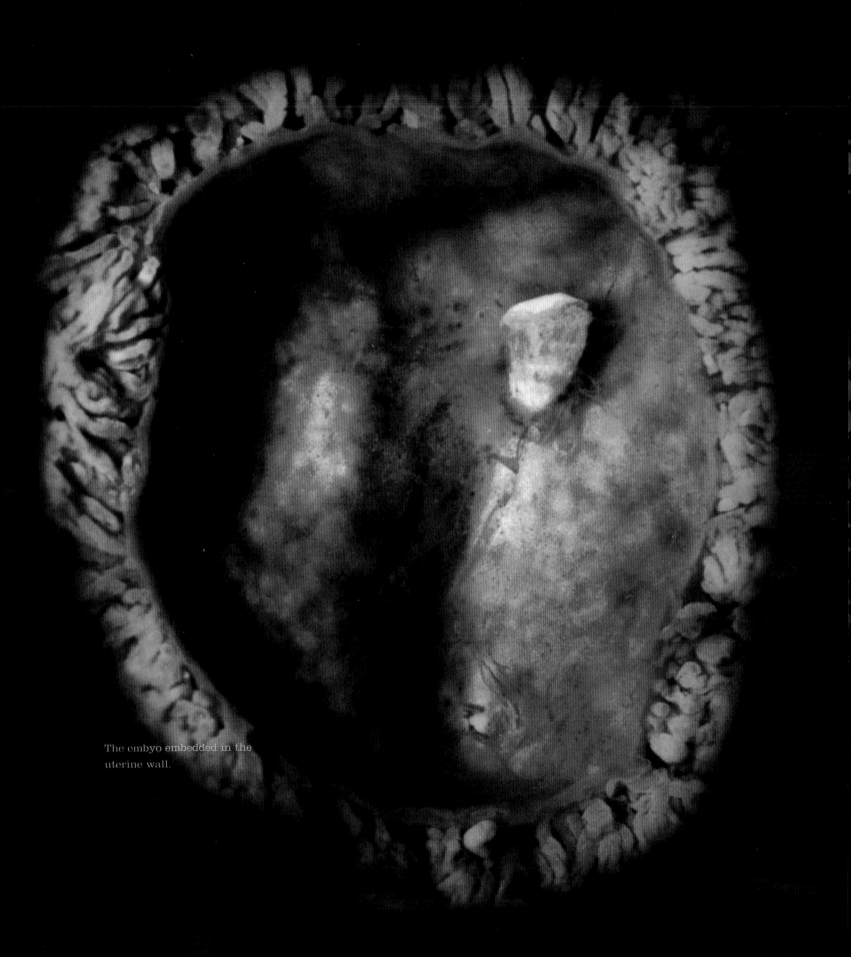

The embyo embedded in the
uterine wall.

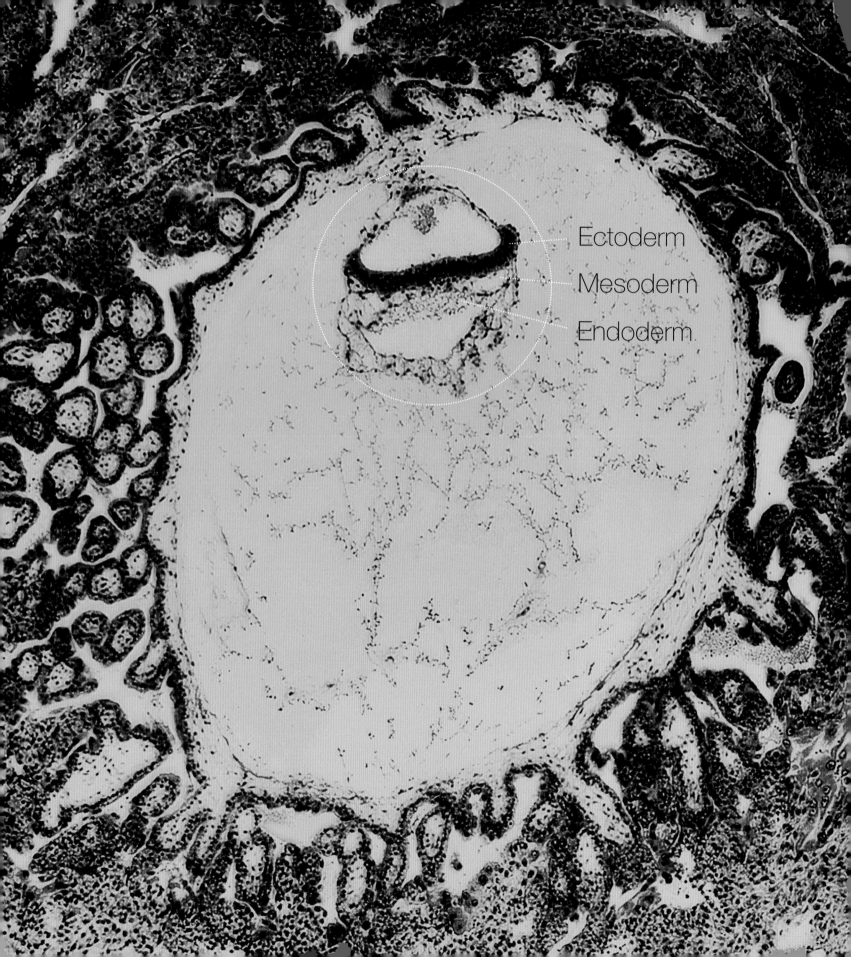

THREE LAYERS

HAVING TAPPED INTO the maternal blood supply and established primitive blood cells and blood vessels of its own, the embryo gluts itself, growing at an astonishing rate. Each day for the next several days it doubles in size — a rate that, if sustained, would cause the baby to be larger than the sun at birth. To sustain this growth, a stalk forms, connecting the embryo directly to the placenta. This becomes the umbilical cord, which eventually, just before birth, will ferry up to 300 quarts of blood per day between mother and fetus.

As the embryo gorges on its first real meals, the inner disc of cells that appeared late the first week resumes its complex unfolding. In nature, function follows shape. Active shapes arise by recasting simpler ones, and by division and subdivision. Thus the soap bubble heap of the morula becomes the hollow sphere of the blastocyst; then the embryonic disc — those layered cells congealing at the upper end of the blastocyst — splits into two layers, endoderm and ectoderm. Each division is a branch point, yielding new possibilities previously beyond reach.

As the disc thickens, a narrow line of cells, the so-called "primitive streak," appears on the surface. So far the blastocyst has given no indication of what it is to become, but by laying down an axis it takes a crucial first step. The line of cells in effect gives the disc an orientation — top, bottom, back, front. Also sides. As cells begin organizing into tissues, then organs and systems, the primitive streak becomes their polestar and guide.

Almost at once, the primitive streak begins to extrude an intermediate layer of cells — the mesoderm — between the disc's original two layers. Together, the three pancakelike cell regions contain the beginnings of a complete and functional human being.

By now the mother may be begin to suspect she is pregnant. Normally, menstruation would begin. When it doesn't, she is likely to want her pregnancy confirmed.

13 DAYS

STOMACH, MOUTH, SKELETON

WITHIN THE EMBRYO are skeleton-forming cells, gut-forming cells—cells of different descriptions. Alive, they begin to pulse, jostle for position, move around. The cells that will later lay down the skeleton travel outward while the cells that will later lay down the gut migrate inward. How they know where to go and align themselves with their neighbors remains a mystery, although scientists have lately begun to understand that certain cells release chemicals that incite others to specific action. The identification of these substances—called morphogens—and the patterns by which they work holds promising implications for understanding not only how we are made but for disease research.

Gastrulation—the process in the development of all animals in which the cells of the blastocyst rearrange and move so that simple flat or spherical shapes transform into something else, something from which the animal can grow—begins not (as the word implies) with the formation of the gut. It starts rather with the cells that will lay down the skeleton. By finding stable contacts with other cells, about 40 of these cells take up a ringlike pattern on the top layer of the embryonic disc. Only after those cells have positioned themselves along the embryo's outer wall, forming a dent, do the group of cells that will become the gut begin to migrate, deepening the depression into a well-developed inward bulge. The once saucer-shaped disc first becomes pear-shaped, broader at one end—the head region. Rapidly, within hours, it resembles an infinitesimal horse collar, with a deep groove halfway up the middle.

Cells at the tip of the groove shoot out long filaments that grab cells across the gulf and pull sheets of them across to enclose the cavity, forming a tube. This is the region of the future mouth. Rapidly, like a closing zipper, the tube lengthens, until the whole primitive digestive tract, from mouth to anus, is fused.

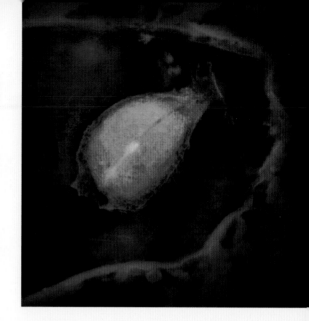

CELLS TALKING

CHEMICAL SIGNALS EMANATING from the mesoderm have so far segregated the production of new cells by functions. Cells massing at the upper level are long and stringy—columnar; cells on the lower are blocklike, almost cubic. Now, the embryo begins to form its third—and most commanding—set of structures.

Many of these outer columnar cells are so-called *neuroblasts*—the earliest cells of the brain and nervous system. These cells cannot carry messages at 300 miles an hour, as mature nerve cells can, but they contain the information for making the cells that make the cells that do. Their progeny will permeate every remote tip of your baby's body.

During the next couple of days, a second sheath of cells pinches and fuses in front of the groovelike primitive streak—the neural tube. Less than 3 weeks since fertilization, the embryo—about the size of a pinhead—has formed the wisp of an apparatus for integrating the actions of the other systems, otherwise known as the central nervous system.

The changes come just in time, as the cells of the embryo have formed channels, and the channels have rapidly formed a network of their own. The task of sustaining and orchestrating more and more complicated cellular developments will soon require far greater coordination than the chemical surges that have so far been in control.

What will become the brain begins to separate. Note the white line where the division will occur. This signals the development of the left and right sides of the body.

16-19 DAYS

Left side

Right side

Left and right

A concave area known as the neural groove is formed. This groove is the first step toward the embryo's nervous system, which is one of the first formations to develop. Right: The blood cells of the embryo are already developed and they begin to form channels.

SOMITES

IF YOU COULD VIEW THE EMBRYO from above, it would look as if it suddenly had been stretched, to resemble the sole of a clown's shoe. The head end is almost double the width of the tail end, the insole slightly tapered. Every ridge, bump, and recess indicates brand new cells assembling into new and different kinds of tissues.

Three pairs of bumps protrude like zipper's teeth on either side of the neural groove, progressing from the tail to the middle. These are somites, dense clusters of midlayer cells that will give rise to the skeleton and the major muscles of the body. A head-fold rises on either side of the primitive streak, which now extends nearly one-third the length of the embryo. In the yolk sac, 2 new types of cells arise: stem cells, which can produce all other types of blood cells, and cells that will form the smooth inner lining of the blood vessels. (Secondary blood vessels also begin to appear at this time in the placenta.)

Meanwhile, muscle cells formed in the topmost layer stream through the primitive streak to a central point just below the head end and assemble there. Like all muscle cells, they are stringy and resilient. Yet those passing through the area nearest the crown create an outward pushing motion when they contract, while those that extrude through the posterior end act more like a siphon. The 2 types of cells rapidly form 2 unconnected crescent-shaped tracts.

Right: The elongated embryo with distinct bump of the brain and the beginnings of the heart.

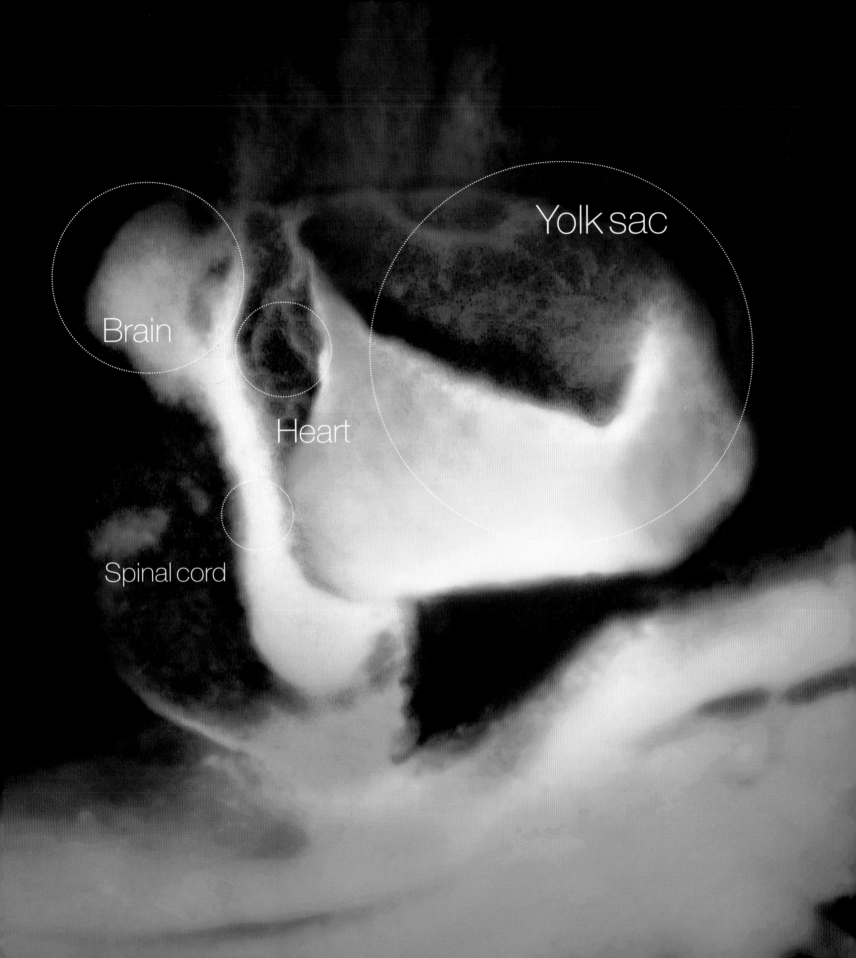

A HEART BEATS

Right: Neural crest cells will eventually contribute to the skull and face of the embryo. Note the clearly defined somites—like zippers.
Top left: An S-shape forms in the heart, and cardiac muscle contraction begins.
Bottom left: At 28 days, the heart is beating.

THOUGH THE EMBRYO is still smaller than the letter "A" on this page, developments quickly add up. Alone, each results from a simple division—a group of cells makes daughter cells that, instead of being exact duplicates, vary slightly in terms of which of their genes are switched on or off. Then their progeny—and their progeny, and theirs—do the same. It's basic stuff, like trees branching. Yet with astonishing speed—out of nowhere, it seems—more and more new and complex structures arise, each more sophisticated and refined than the last.

As the embryo becomes longer and the yolk sac expands, more somites appear—by now, up to 12 pairs—suggesting the beginning of a spinal column. As they emerge, they enclose corresponding portions of the neural tube, encasing the area where the spinal cord will be strung. Just above the top pair, on either side of the neural groove, the cells that will become the eyes appear as thickened discs. The cells of the ears are also present, although they have yet to produce any corresponding new structures. Similarly, so-called *neural crest cells* accumulate on the underside of the crown—there eventually to bulge into skull and face.

Most remarkably, the two crescent-shaped tracts in the upper midregion fuse into a singular S-shaped tube, activating the upper tract to start pumping while the lower begins the reverse cycle—suctioning. Barely 3 weeks after fertilization, with a primitive vascular system unspooling by the hour, a human heart begins to beat.

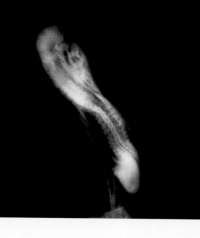

A primitive S-shaped tubal heart is beating and the rhythmic flow of fluids throughout the body begins. However, this is not true circulation because blood vessel development is still incomplete.

➤ Facing page: 25-27 days postovulation. At this stage, the brain and spinal cord together are the largest and most compact tissue of the embryo. A blood system continues to develop. Blood cells follow the surface of the yolk sac where they originate, move along the central nervous system, and move in the maternal blood system.

THE MOTHER KNOWS

BY NOW, MOST EXPECTANT WOMEN know they are pregnant, alerted by a menstrual period that hasn't arrived, or else hormonal changes that leave them feeling agitated and nauseous. Typical of the maternal-embryonic bond, the more unsettled the mother feels, the more vibrant the upheaval occurring inside her.

As the end of the first month approaches, the embryo assumes a distinctly vertebrate shape — bulbous head, arching back, elongated tail. In less than 4 weeks, it has gone from one cell to millions, from an unformed mass of protoplasm to intricately organized groups of cells that form the basis of most of the body's major systems — nervous, muscular, vascular (circulatory), digestive, and skeletal. The foundations of the heart, brain, spinal cord, and sense organs have been laid down and are functioning, although the organs themselves are far from formed. The tubal heart beats with regularity, propelling fluids throughout the body, but the result is not true circulation, since most blood vessels have yet to be formed.

As the remaining pairs of somites are laid down, the neural tube closes. The incipient brain and spinal cord are now the most compact and powerful tissue in the body and begin to integrate the other systems. Meanwhile, the blood system continues to develop: Embryonic blood cells follow the surface of the yolk sac, where they are formed, move along routes of the central nervous system, and travel to the placenta and back again. Within days, cells lining the gastric tube will begin to differentiate into several new regions — liver, lung, stomach, and pancreas.

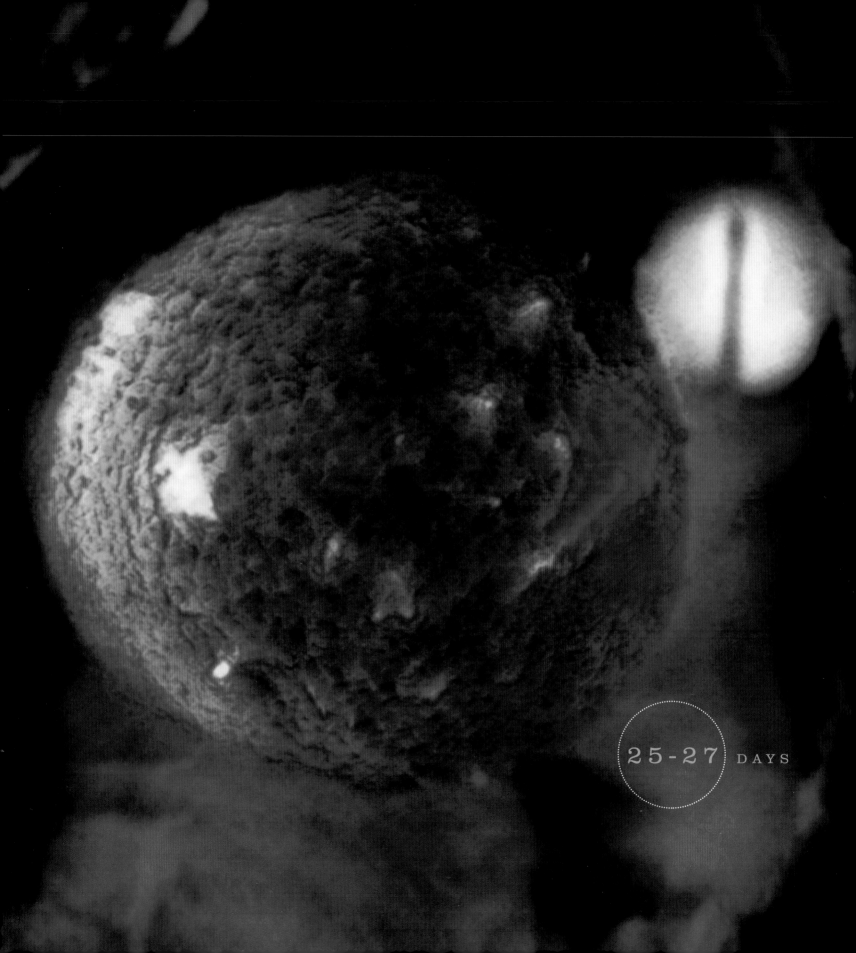

25-27 DAYS

Back view

Back view showing the division between the
inner and outer cavities of the embryo.

23-25 DAYS

Front view

Left: The central nervous system appears to dominate the whole embryo.
Below: The tail close up.

By the time the neural tube is closed, both the eye and ear will have begun to form.

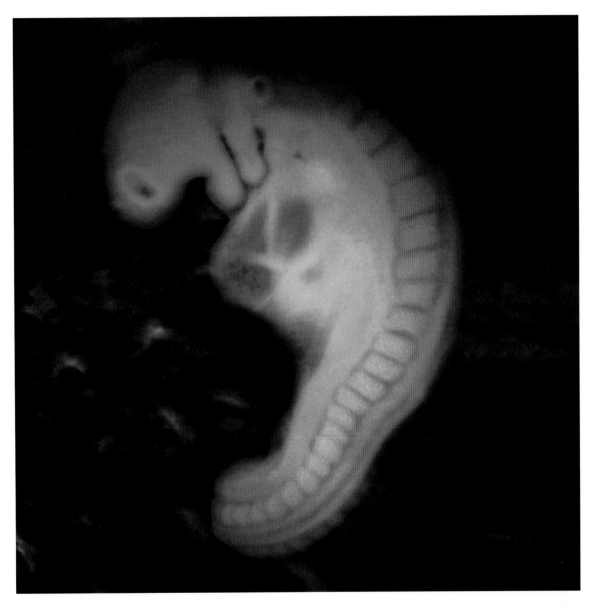

The embryo curves into a C-shape. The arches that form the face and neck are now becoming evident under the enlarging forebrain.

A Miracle Every Day

ORCHESTRATION

GROWTH

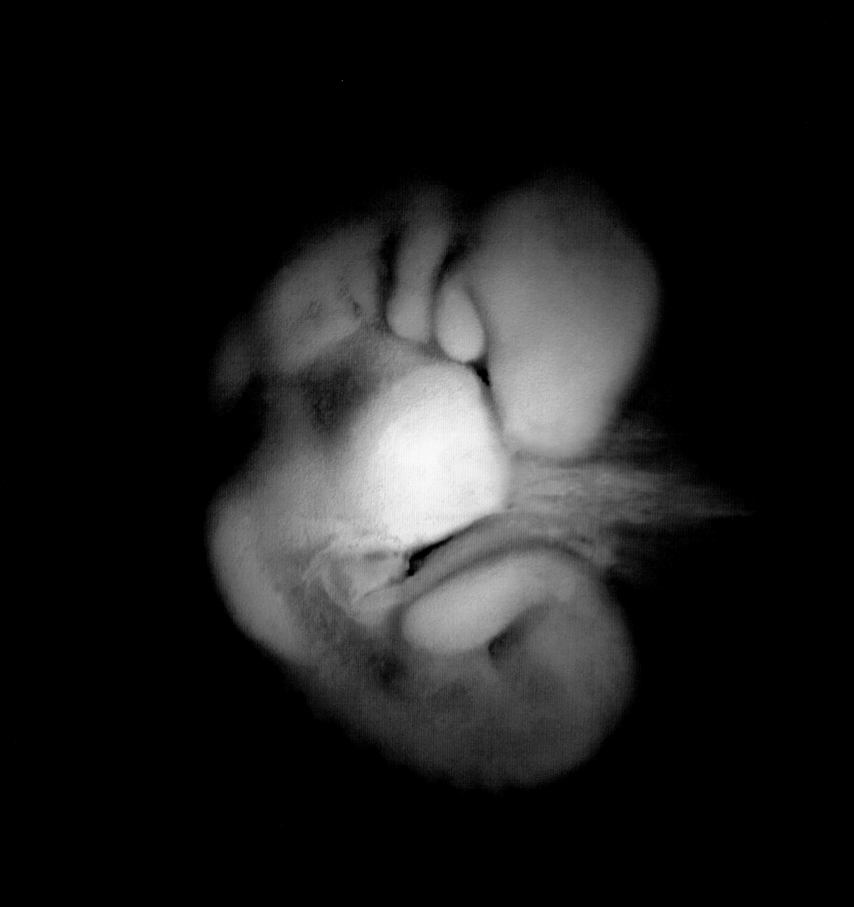

03

Metamorphosis

"Every mom-to-be," the writer Katherine Whittemore notes, "evolves into a conversation piece." Every dad-to-be evolves into a lamp shade. There is no injustice here, for it is in the woman that the greatest developments all take place. The embryo is transforming virtually day by day — and with it the mother's role and identification. Though her pregnancy does not yet show, she knows she is carrying a child, and that changes everything.

As the second month begins, the emerging embryo is poised midway in the period of its greatest growth and development. Four weeks earlier, it was a single cell; 4 weeks hence all its vital parts (with a few exceptions) will exist and it will look unmistakably human, if not yet like its parents.

Straddling such a threshold, the embryo at this stage has long been a mystery. Side by side, it could easily be confused with the offspring of other species. Mice, elephants, pigs, even chicks look almost identical at the same stage. More, it retains ancestral structures

—a tail, the yolk sac, the beginnings of gills—that evoke lower creatures and that it will soon lose.

A century ago scientists theorized, based on these resemblances, that we each pass through all the stages of evolutionary development en route to becoming who we are. That notion is now widely discredited—we do not go through a fish phase, a bird phase, a reptilian phase, an ape phase, before we become human. But human origami, however magnificent, is still origami. The processes of development are constant; it's how they're used that results in what Darwin called "the most exalted object which we are capable of conceiving"—a human life.

What marks the second month of pregnancy is not the change in the embryo's appearance, even though, in pictures, this is when we first witness its thrilling humanity and fall in love with what we see. The true significance is in the degree of completeness. In a single month, the hint of heart becomes a heart, the shadow of an eye becomes an eye, a tube of cells twists and bulges into a multiorgan system. Four ridges are sculpted into a human face. Life unfolds at its most magnificent, in less time than to grow a head of lettuce.

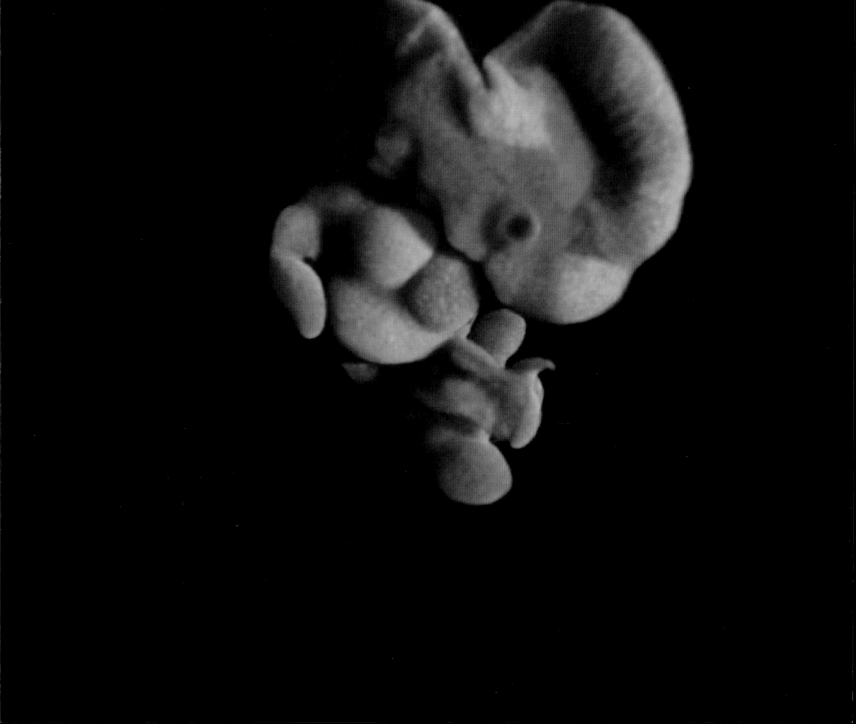

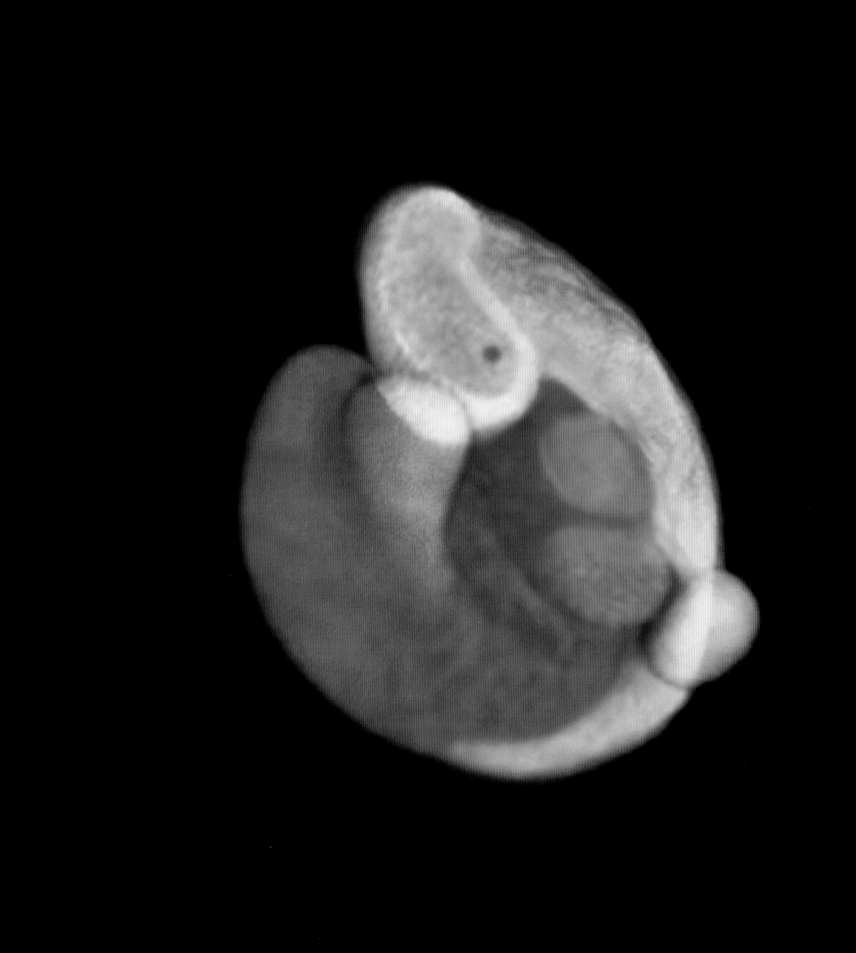

AN EXPLOSION OF CHANGES

SYSTEMS DEVELOP: Growing from within, the embryo at 28 days curls in a characteristic C-shape, a tight comma. Its primitive heart and brain are active, in place and folding rapidly into complex and more functional shapes. The top is enlarged, like the hasp of a safety pin, as the brain region thickens and begins organizing into sub-regions. Although the embryo is still relatively featureless, a system of ridges — precursors of the face and neck — arch inward from the underside of the crown. Each ridge is already wired with a major artery and a cranial nerve — trunk lines, so to speak — and contains a rod of cells designed to grow into bone. Contained within the three arches are the essential apparatus for hearing, speaking, breathing, eating, and facial expression. Cell plates thickening along the top of the neural tube near the cranial region form the beginning of the eye.

Meanwhile, the squiggly heart tube has developed "prominences" — bulges that auger the formations to come. This is the great planning period, and the building blocks for 40 pairs of muscles and 33 pairs of vertebrae radiate out like hard fibers from the spinal column. Another prominence — the mesonephic — resembling primitive eel kidneys filters metabolic wastes from the blood, standing in temporarily for the urogenital system. Flipperlike limb buds, upper and lower, extend from the bottom half of the C, which tapers to a spiral tail. No bigger than a grain of rice, the embryo floats like a coiled seahorse in a tiny saline sea, tethered by a hair-thin cable of twisting arteries inserted into the placenta, a satellite organ where food, oxygen, and waste are exchanged with the mother.

◄ Facing page: The liver and heart combined are equal in volume; the early division of what will be the heart's 4 chambers are now clearly visible.

Embryo at 28 Days

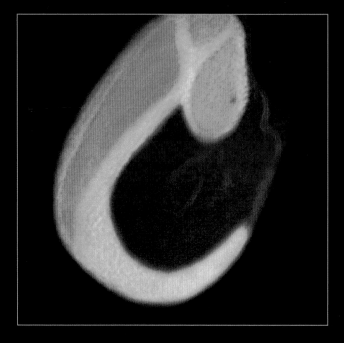

ACTUAL SIZE 4.0 MM

A VIEW INSIDE THE EMBRYO AT 28 DAYS
The pace of development is astonishingly rapid at this
point—at 3 weeks, very little has developed; by 8 weeks,
all bodily systems are in place.

Facial cavities

Eye lens

Heart

Nose

Tail

Arm bud

Somites

Leg bud

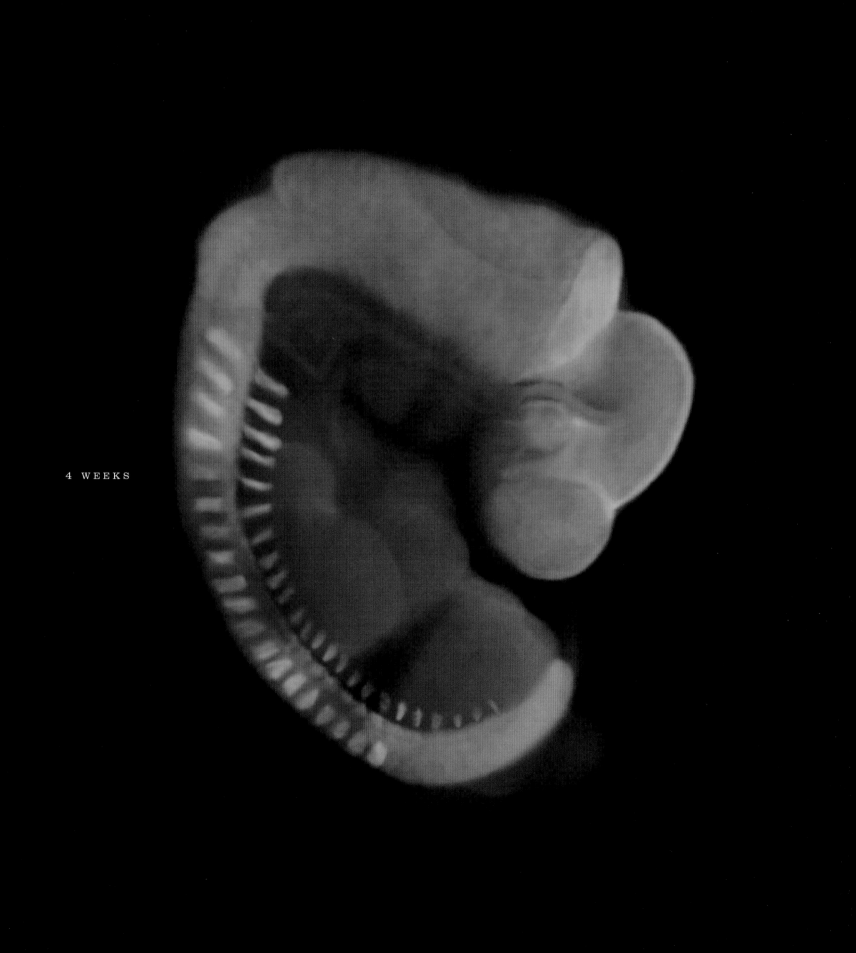

4 WEEKS

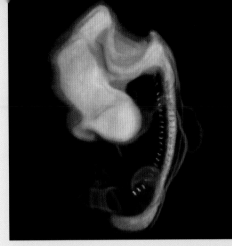

6 WEEKS

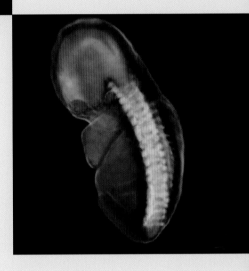

14 WEEKS

NERVOUS SYSTEM

COMMUNICATION: The human nervous system has been called the world's most efficient network for transmitting messages. It begins to form at 18 days gestation and continues to develop until several weeks after birth, penetrating to the minutest regions of the body. Conveying information along a trunk and branch system of intertwining cells at jet speed, it connects the brain with nerve endings in the muscles, glands, and sense organs. At birth 10,000 taste buds in the mouth, 240,000 hearing units in the ears, and up to 50 billion light-sensitive points in the eye will deluge the system with data, as the baby begins to see, hear, and taste the world.

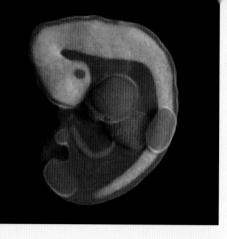

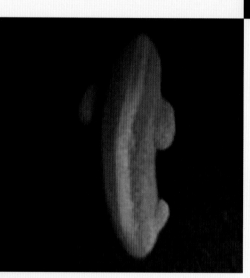

Above right: The heart is divided into 4 subsections.
Above left: A back view showing the limb buds.

➤ Facing page: Photograph from the back showing the somites.

BECOMING...

DISTINCTION: Like a photographic image clarifying in a developing bath, the embryo grows more distinct each day as its finer features emerge. An arched groove — the hint of an ear — appears above the ridges in the face region. The lens pits deepen, as the remarkable process of building a pair of eyes begins. Two thickened cell plates from the original outer layer of cells expand, then develop craterlike depressions surrounded by horseshoe-shaped rings of cells; these will become the nose. A gill-like feature — the lateral cervical sinus — appears along the inmost facial ridge, eventually to be overgrown with skin. Paddlelike limb buds elongate.

Inside, the heart and brain rapidly expand and take shape. Both organs are ventricular — they consist of central canals that expand into tributaries connected by thinner channels and are filled with fluid. Within days after the brain starts to subdivide, a new ventricle opens containing the primitive ductwork needed for motor control and sensations.

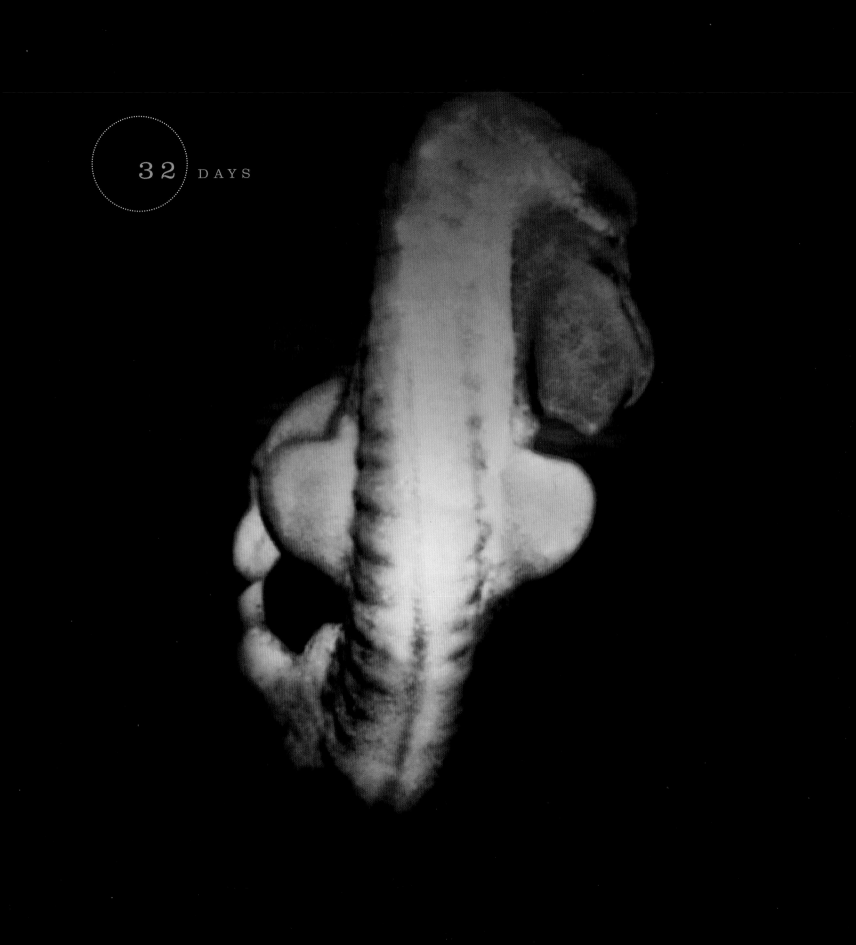

Embryo at 32 Days

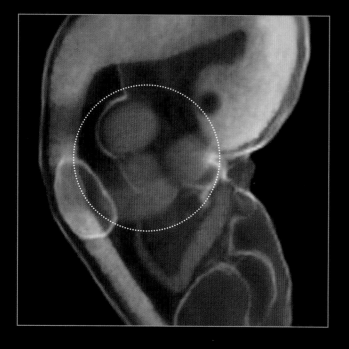

ACTUAL SIZE 4.5 MM

THE HEART

The 4 chambers of the heart have developed. The right
ventricle and right atrium are clearly visible.

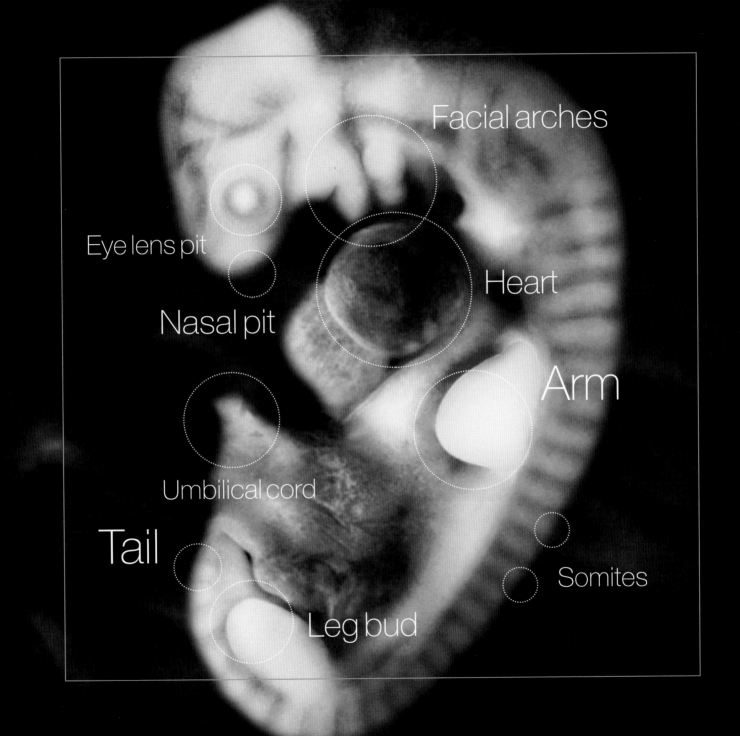

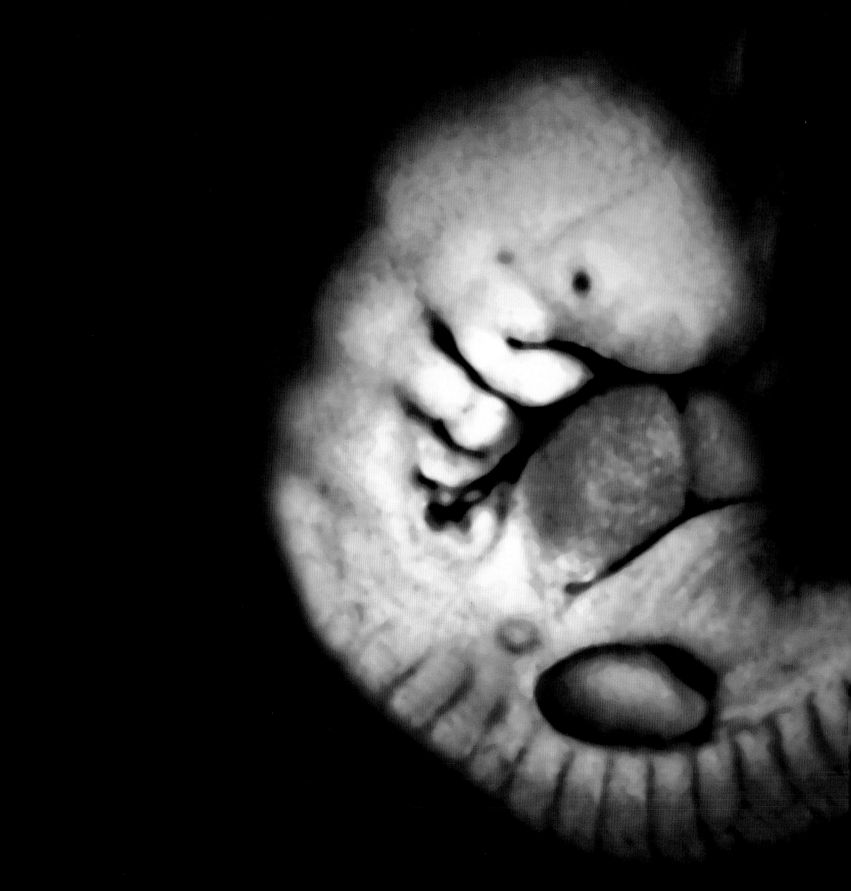

The computer and the motor

Fifty percent of the embryo is brain and heart.

8 WEEKS

ADULT

> Facing page: The heart has gone
through its most dramatic structural development by the end
of the 8 weeks. The fetus is about
one-and-a-half inches tall at this
stage. Already the resemblance to
an adult heart is startling.

THE HEART

Developing from the third week on, the embryonic heart is built like a transformer, one
of those toys that morph into something else, say, a rocket that unfolds and refolds
into a superhero. Throughout pregnancy, it needs only to act like a single pump, main-
taining blood flow throughout the body and into the placenta. Yet to meet the circulation requirements of life after birth, when oxygen molecules must be grabbed from
the air through breathing, the embryonic heart must also form 4 chambers that can
alternately pump and receive blood from the lungs. This physiological dilemma is
solved by the presence of 2 shunts that allow each chamber of the heart to handle
large amounts of blood while sparing the underdeveloped vessels in the lungs. These
shunts automatically are sealed shut at birth, as the baby breathes — and cries — for
the first time.

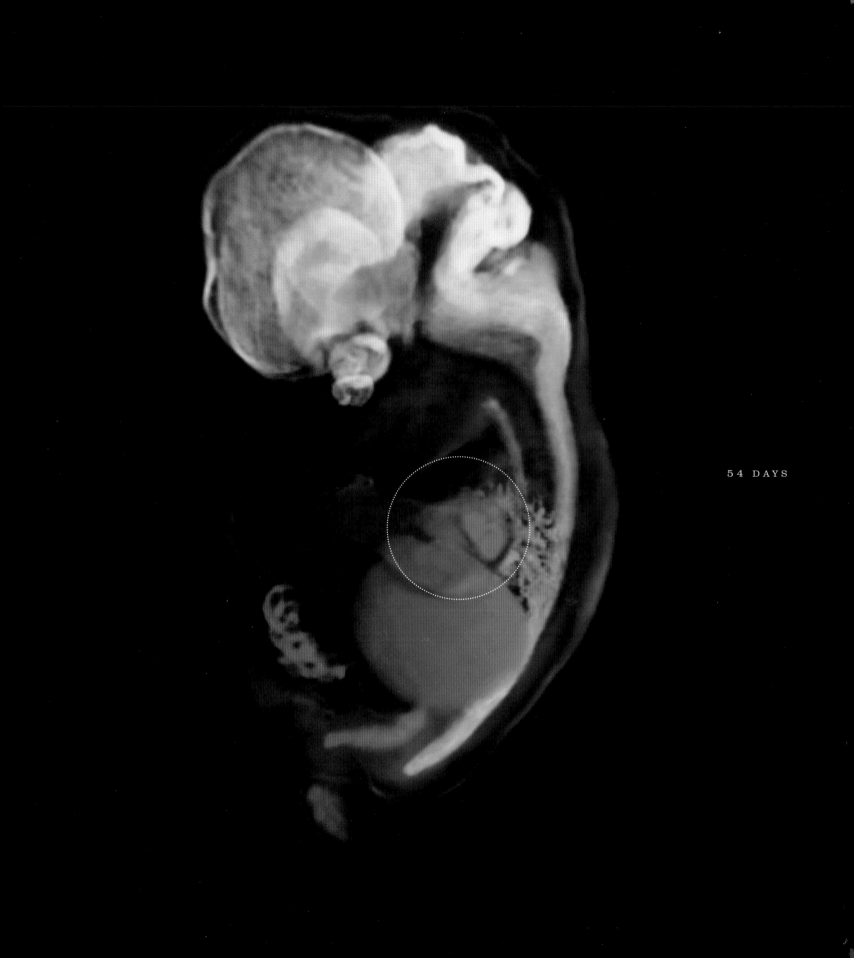

54 DAYS

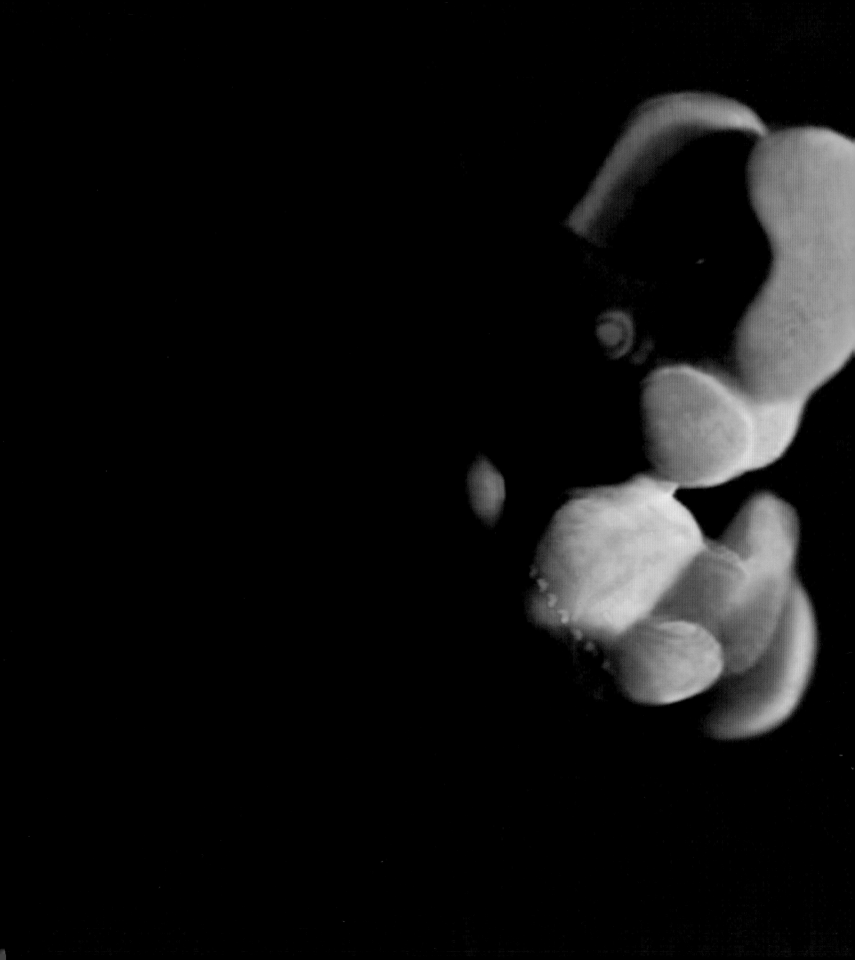

ORIGAMI

UNFOLDING: With remarkable speed, the origami of human form—making accelerates, yielding more and more specialized shapes. The brain is now a series of 5 cavities lined with embryonic nerve tissue, an infinitesimal chain of caverns. Cells near the crown start to differentiate into those that will fill the cavern furthest from the heart—the cerebral cortex, seat of the intellect. Ear pits expand and take shape. The nasal pit appears to deepen and climb toward the eye.

Since the heart and brain so far have been needed sooner than the digestive organs, the lower half of the body has developed more slowly than the upper. Now, though, the liver, stomach, and esophagus start to form. The upper limb buds segment, suggesting shoulders, wrists, and hand plates, while the lower buds flatten, flipperlike. Cells for determining if the baby is a boy or girl migrate from the yolk sac toward the site where the sex organs will develop. The vestigial tail reaches its maximum length before starting to recede and vanish.

The embryo remains bent double, so that the head and tail nearly touch, just as with other vertebrates at this stage. Outwardly, it looks scarcely different than a chick, mouse, or pig embryo. But inside the apparatus for making a distinctly *human* individual is now fully active, and the cumulative effects have started to show.

Left: Front view of the embryo. The brain has increased its size by 30 percent in the last 4 days.

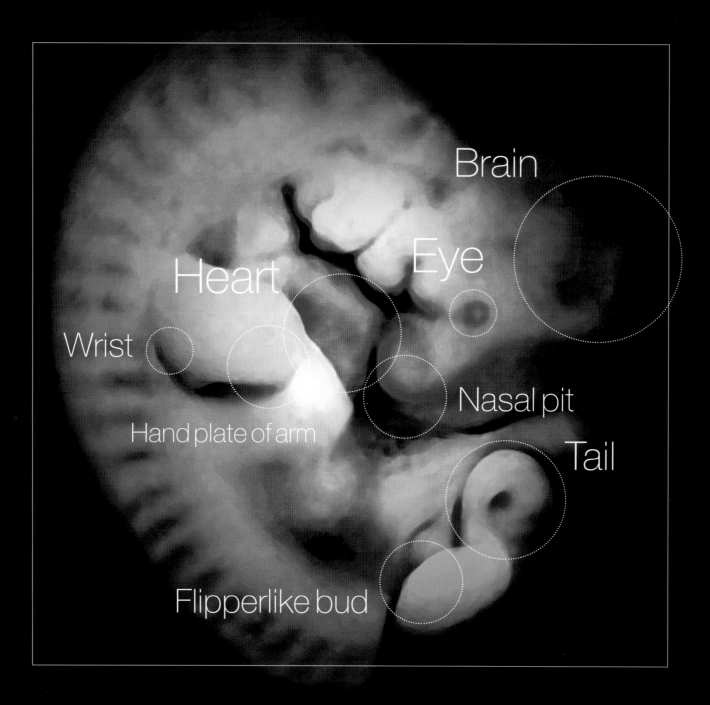

Lıııılıııılıııılıııılıııılıııılıııl
ACTUAL SIZE 6.0 MM

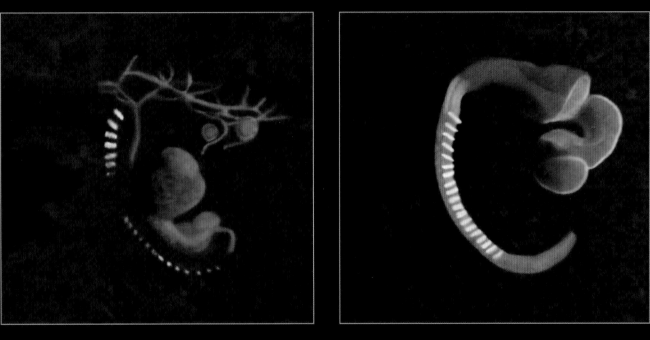

PARTS OF THE WHOLE

The imaging technology that makes this book possible
allows us to see marvels that have never been viewed this
way before. Here we can see the eye, the circulatory system feeding the brain, the heart, liver, the beginning of

Nervous system with the brain and nerve endings.

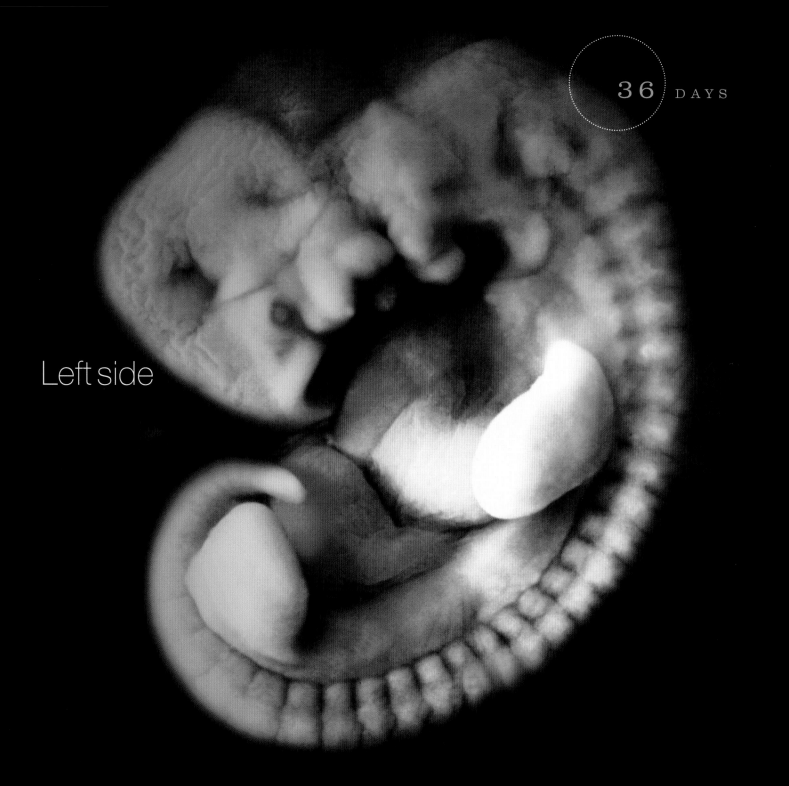

Left side

Right side

The external ear becomes visible.
Leg bud begins to round at top
and the tip of its tapering end will
eventually form the foot.

Above: Close-up of the tip of the
tail—front view.

➤ Facing page: Like looking through
a glass house, making the skin
translucent allows us to see the
cardiovascular system.

THE BABY'S FACE

FEATURES: Daily, new shapes beget newer ones.

The features of the face begin to assemble rapidly. A groove is laid down next to the nasal pit—the incipient lower jaw and lip. A day or two later, the upper and lower jaw parts, which have formed in symmetrical halves separated by a gap, fuse. Two days after that, with the jaws well formed, the precursors of the teeth and facial muscles begin to grow and the vestigial gill arches recede.

Tiny mounds of tissue erupt where the whorl of the external ear will grow. The eyes begin to show color, then within 24 hours eye muscles—among the body's most delicate—begin to form. Nerve fibers connect with the olfactory region of the brain, paving the way for the sense of smell.

Meanwhile, the heart begins to subdivide into pumping chambers while the liver takes up the job of making new blood cells from the yolk sac, which begins to wither.

If the first month was the great planning period, now—the second month—is when the building of the baby reaches its peak. Though risks to fetal development remain, they subside sharply as the embryo evolves into a more distinctly human form.

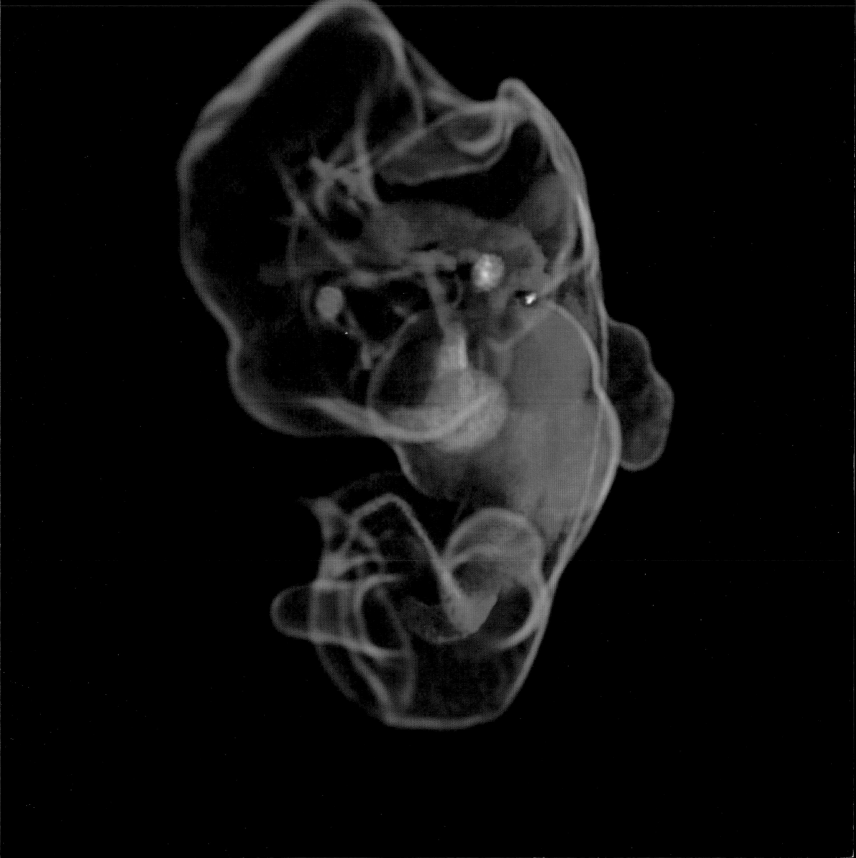

Embryo at 40 Days

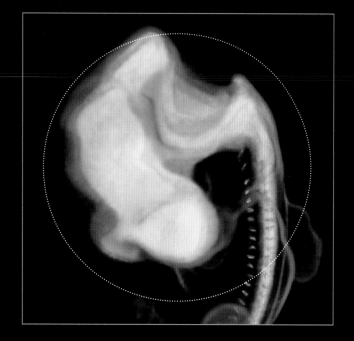

|⊥⊥⊥|⊥⊥⊥|⊥⊥⊥|⊥⊥⊥|⊥⊥⊥|⊥⊥⊥|⊥⊥⊥|

ACTUAL SIZE 8.0 MM

BIG BRAIN

Note what a large percentage of the body the brain takes up.
It is already responsible for regulating breathing, heart,
and muscle movements.

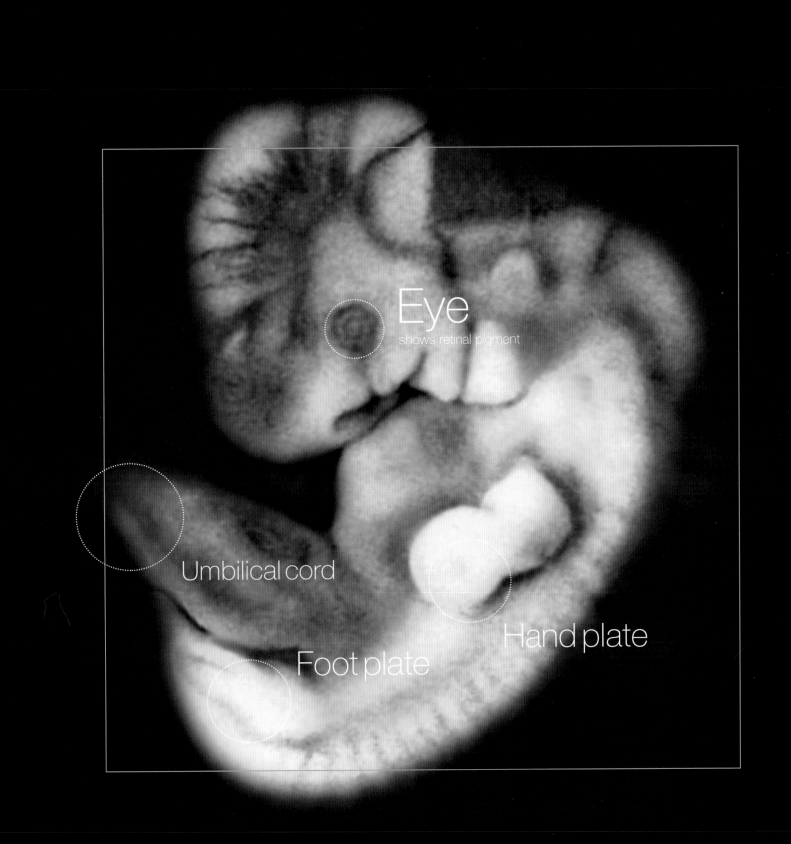

Eye
shows retinal pigment

Umbilical cord

Hand plate

Foot plate

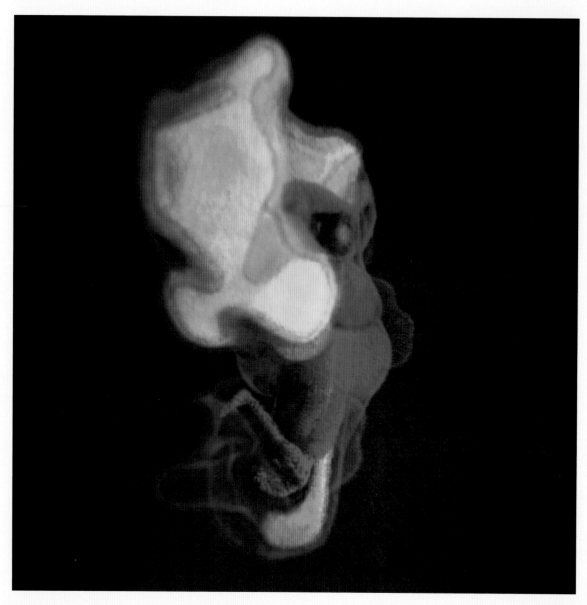

Front view emphasizes the proportions of the nervous system and the blood, highlighted white, compared to the body.

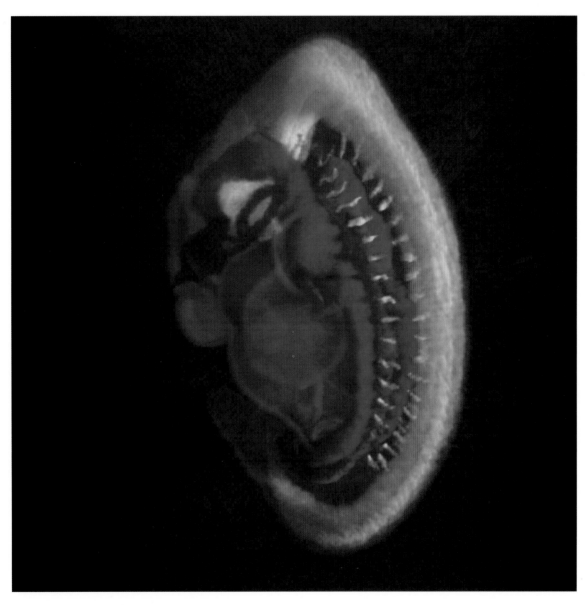

Back view—nervous system with nerve endings.

Arm

Back view: Upper limb develops
into an "arm."

Chambers of the heart

Blood flow

Note the complexity of the blood flow
feeding the embryo.

40 DAYS

Nervous system

LEFT SIDE

BACK VIEW

40 DAYS

Revolutionary views of
a human developing

Embryo at 42 Days

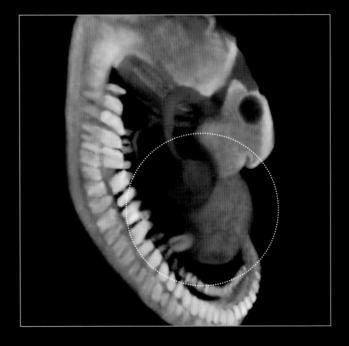

|⊔⊔⊔|⊔⊔⊔|⊔⊔⊔|⊔⊔⊔|⊔⊔⊔|

ACTUAL SIZE 11.0 MM

A VIEW INSIDE THE EMBRYO AT 42 DAYS
Neural buds along the spinal cord are clearly developed.
These tiny buds now are merely one-tenth the width of
a human hair and will grow to more than half an inch in
some cases. Also, at this time, the embryo develops the
sense of smell. Inside the circle one can now see the stomach and liver.

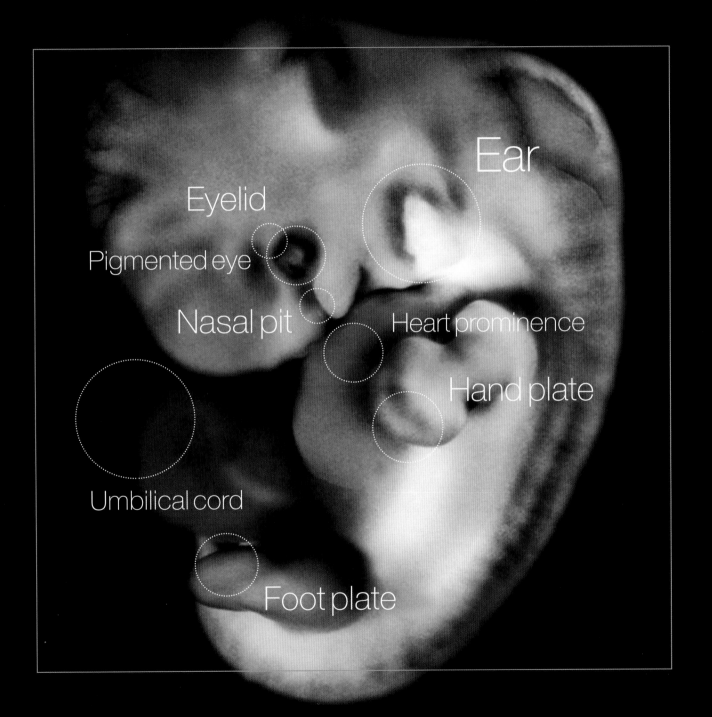

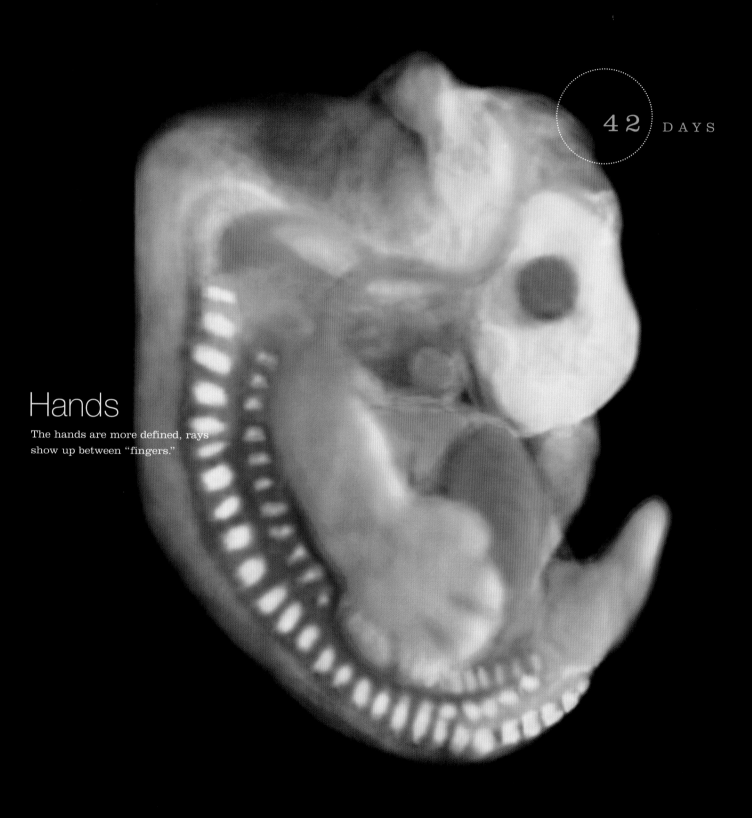

Hands

The hands are more defined, rays
show up between "fingers."

Hormones

In the brain, the gland responsible for
the growth of hormones develops.

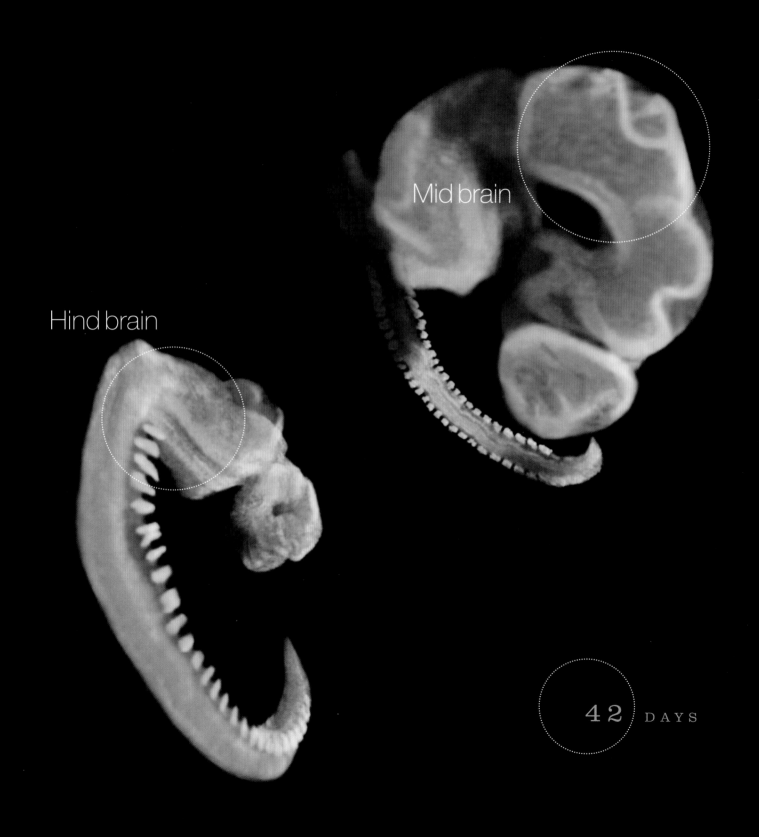

Mid brain

Hind brain

42 DAYS

Forebrain

Seeing, feeling,
understanding

Clear views of the forebrain (what will allow
a person to experience emotion and understand
language), the midbrain (what will allow a
person to hear), the hind brain (what will allow
a person to see), the central nervous system,
and the neural buds.

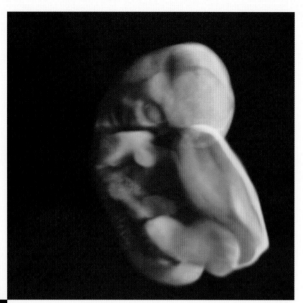

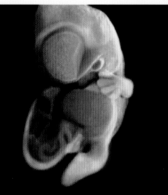

EYES, EARS, AND FINGERS

44 DAYS

THE SENSES: Though still only a half-inch long and weighing less than a paper clip, the embryo sprouts the rudiments of fingers. Nerve cells have formed in the retina. The palate and the semicircular canals of the ear are laid down. Milk lines, which anticipate the production of mammary glands, arise in both sexes. And in the male, the penis begins to form. As cartilage and bone elongate throughout the body, the foot plate becomes well formed and the ankle region is visible. Note in the image on the right that the hands and fingers develop approximately 3 to 4 days ahead of the feet and toes.

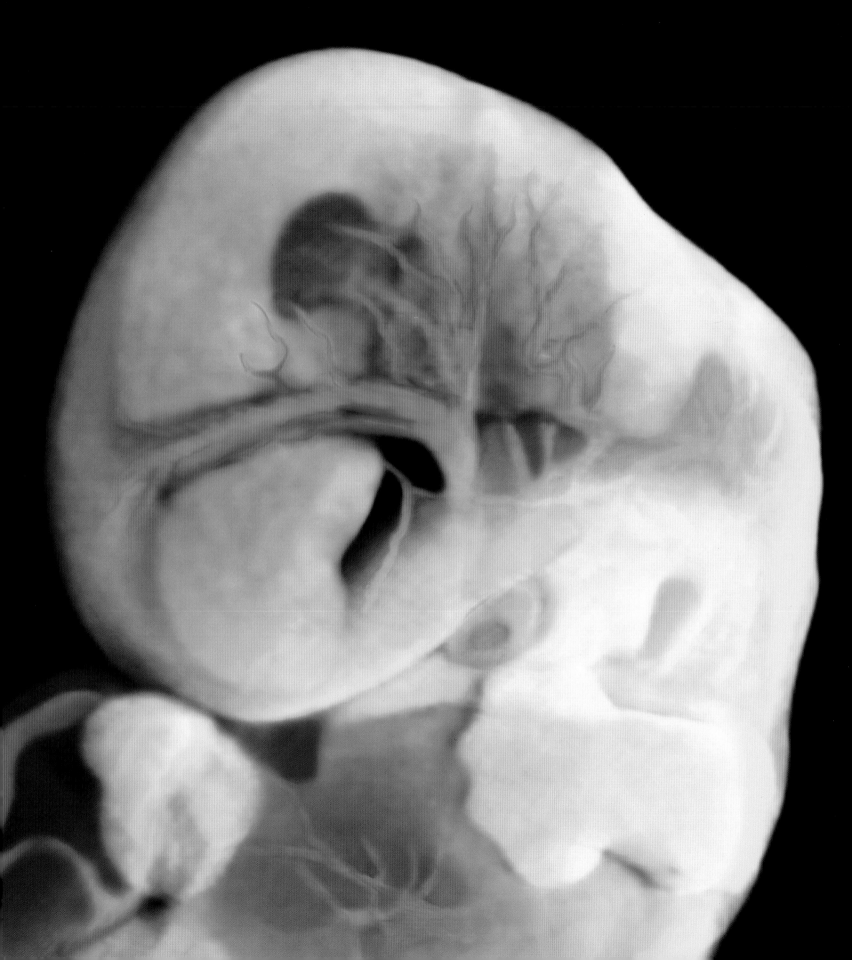

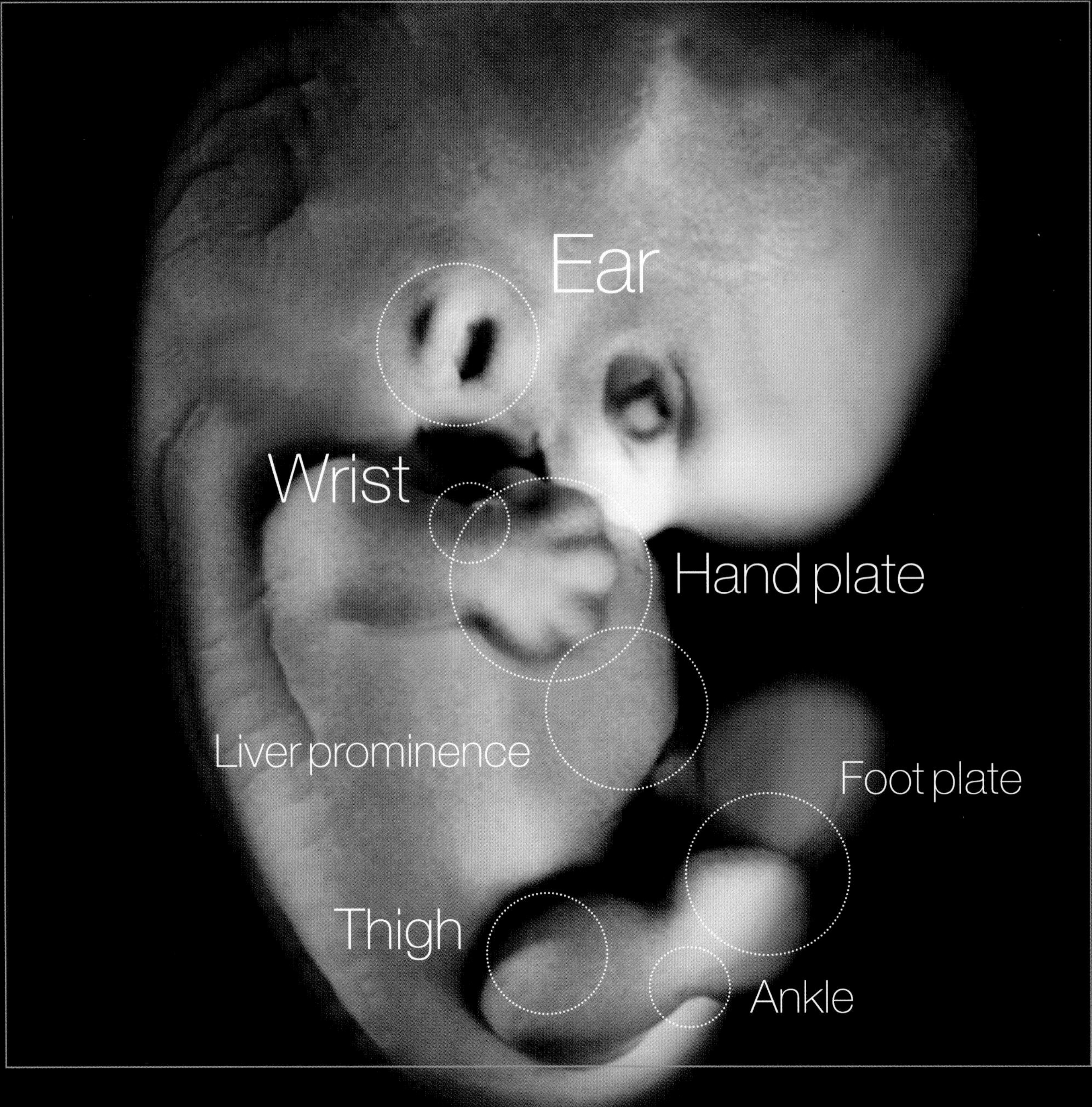
Ear
Wrist
Hand plate
Liver prominence
Foot plate
Thigh
Ankle

Embryo at 44 Days

ACTUAL SIZE 13.0 MM

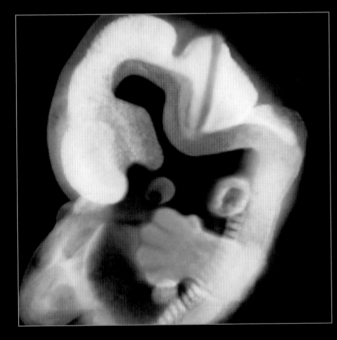

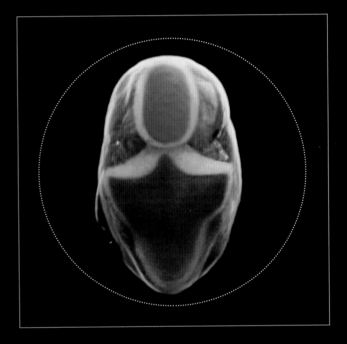

THE BRAIN

Fourth ventricle (white) of the brain. This section controls blood flow.

Top view showing the relationship of the brain (yellow) and the blood flow in the brain (purple). Nerves becoming part of the scalp area.

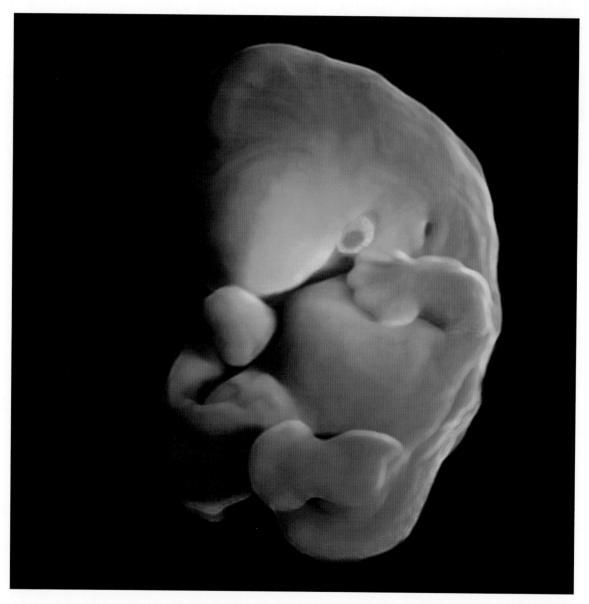

A view of the 44-day-old embryo through the author's visualizations technology.

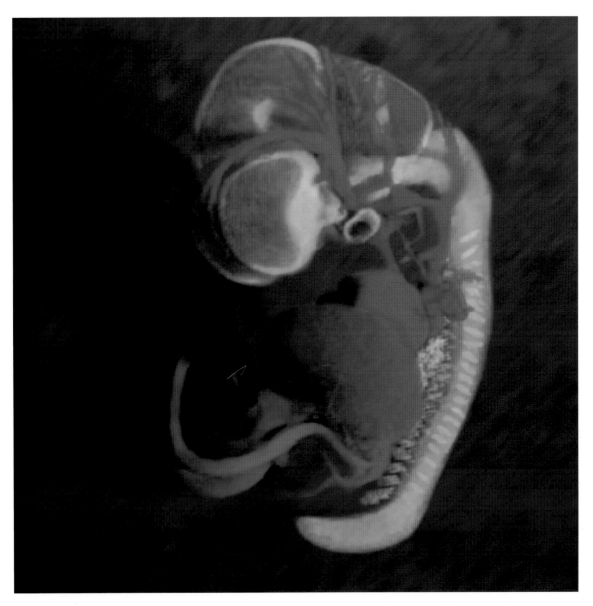

This technology allows you to view the same embryo, but see how well-developed its internal systems are even at 44 days.

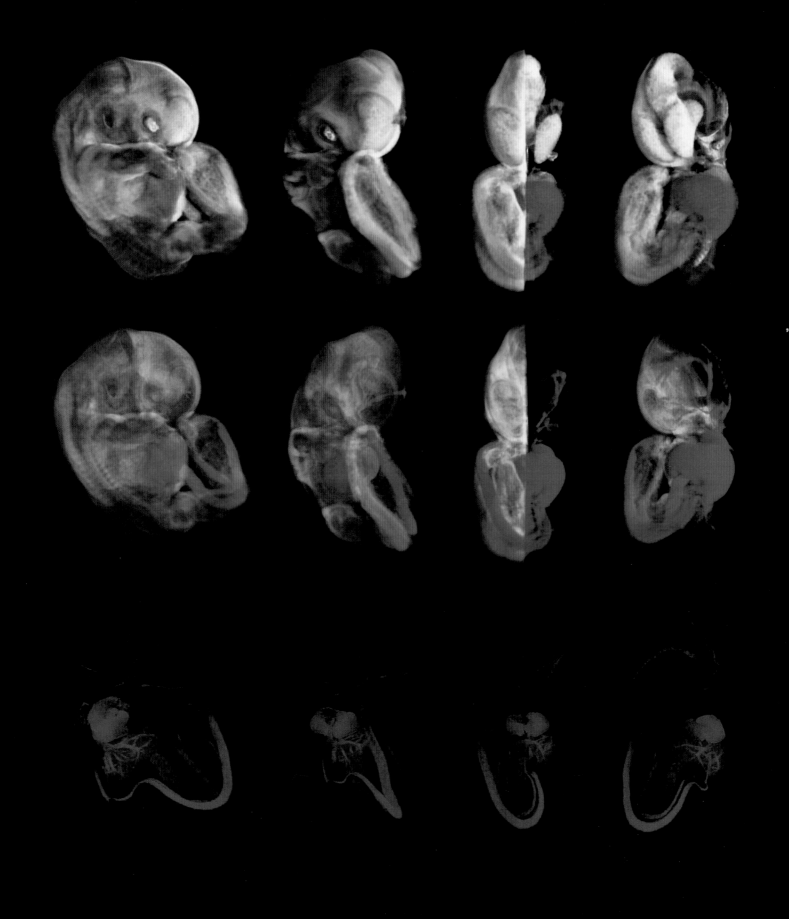

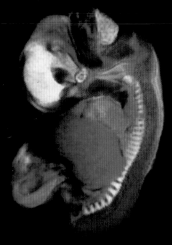
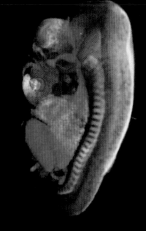
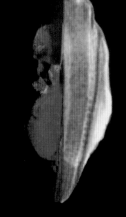
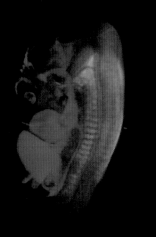

Revealing the blood flow

Through rotation and gradual imaging techniques, these images show the vast amount of blood circulation needed to fuel the extraordinary growth of a 44-day-old fetus. The images at the bottom show only the circulatory system.

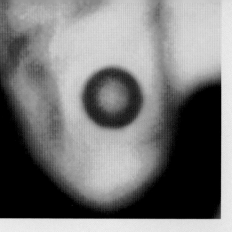

4 WEEKS

THE EYE

Development of the eye is first evident at about 22 days gestation, with the appearance of 2 symmetrical bulges — optic grooves — on the earliest brain-plate tissue. The human eye is an enormously complex structure. The cornea and the lens must be transparent and properly aligned to provide a pathway for light from outside to the retina. In turn, the retina — consisting of millions of light-sensitive cells — must be arranged in such a way as to receive visual signals and relay them to the proper parts of the brain. While all this is developing, during the second trimester, more than one million optic nerve fibers are growing outward from the brain to mesh with each eye. Tunneling through bony sockets, they somehow meet precisely. When they do, at about 6 months, the eyes become sensitive to various levels of light and darkness, but still cannot perceive objects.

6 WEEKS

Development of the eyes is ongoing through the whole pregnancy.
Week 4: A tiny visible dot shows the eye is already present.
Week 6: Eye becomes distinct with a pigmented retina. Eyelids start developing.
Week 8: Developing eyelids start closing up.
Week 24: Eyelids completely close up around 9 weeks and reopen around 26.

8 WEEKS

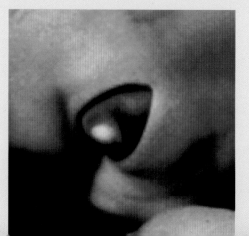

24 WEEKS

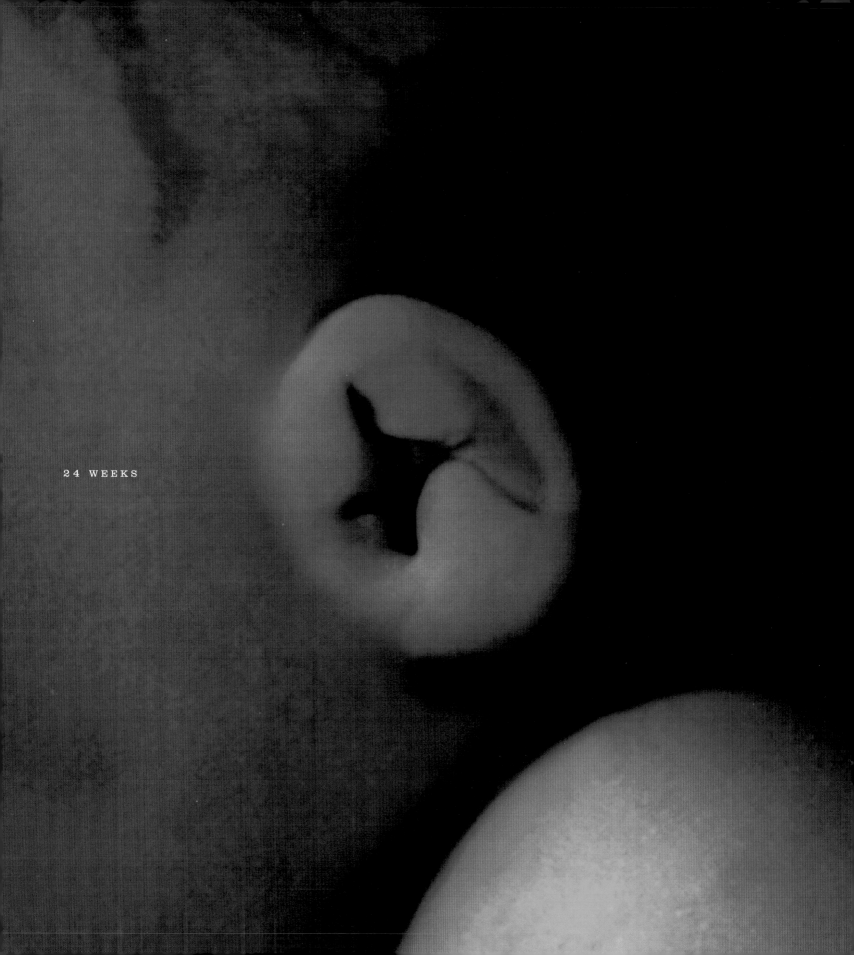

24 WEEKS

4 WEEKS

6 WEEKS

THE EAR

HEARING AT 24 WEEKS: Among the senses, the ability to hear is what first tells us we aren't alone. Pinched, folded, furled, and grooved, the cornucopia-shaped ear is sculpted, gradually, in 3 sections from the embryo's thin skin and a prominence in the lower face. (Eventually it will be sensitive enough to discern whispers and durable enough to withstand cannon bursts.) At between 5 and 6 months, the fetus hears its first sounds—the cacophony of its mother's body (digestive rumblings, rushing blood) and at the far margin of its world, her voice.

As the jaw develops, the ear's position changes significantly.
Week 4: A visible groove appears where the ear will develop.
Week 8: The external ear begins to form.
Week 24: External ears reach their final position around 14 weeks, they stand out from week 16, and continue developing to their final size and shape throughout the rest of the pregnancy.

8 WEEKS

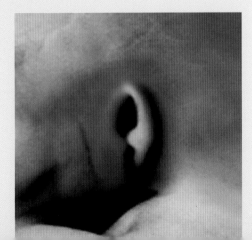

Protected and nourished

The baby in the amniotic sac.

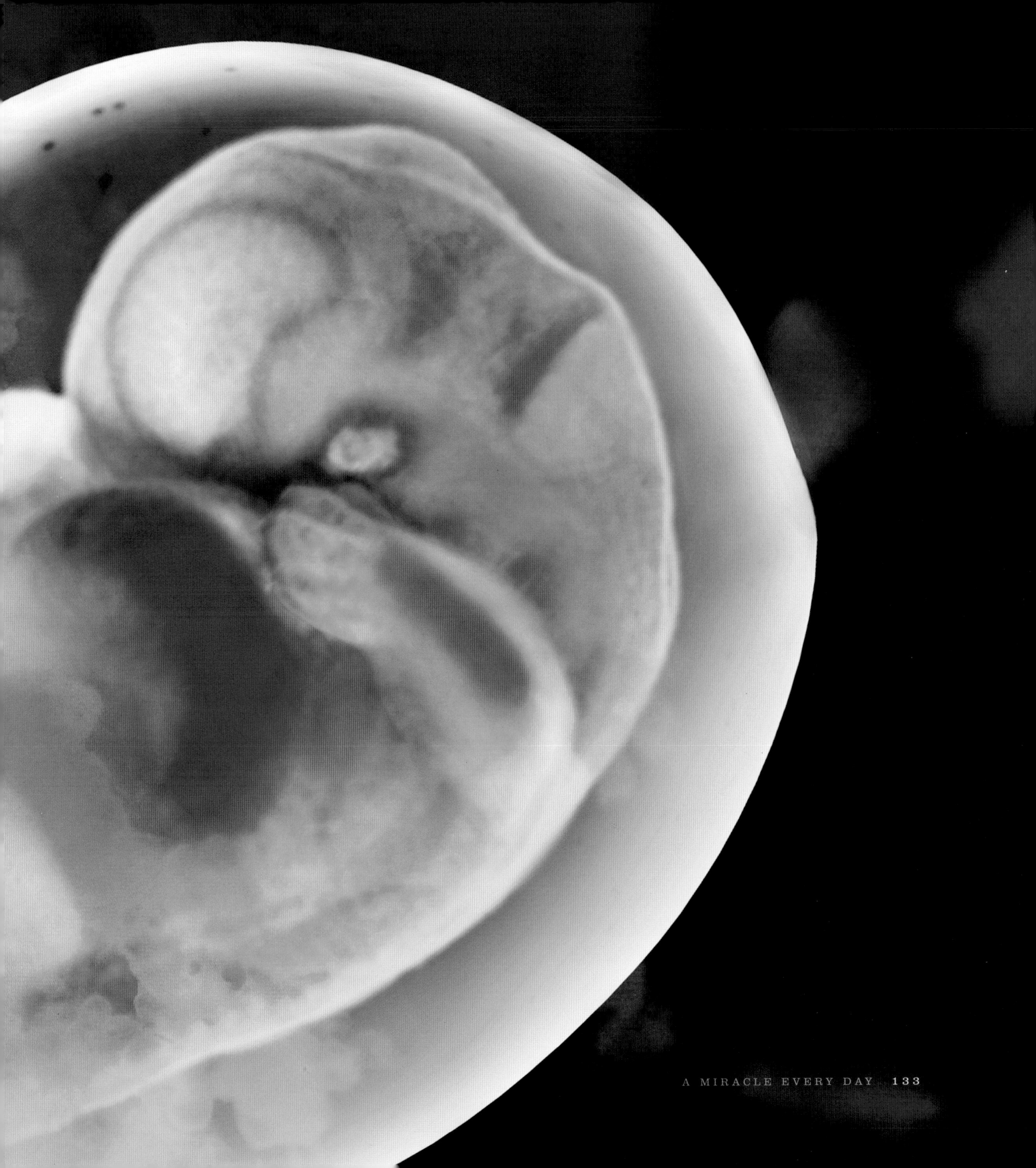

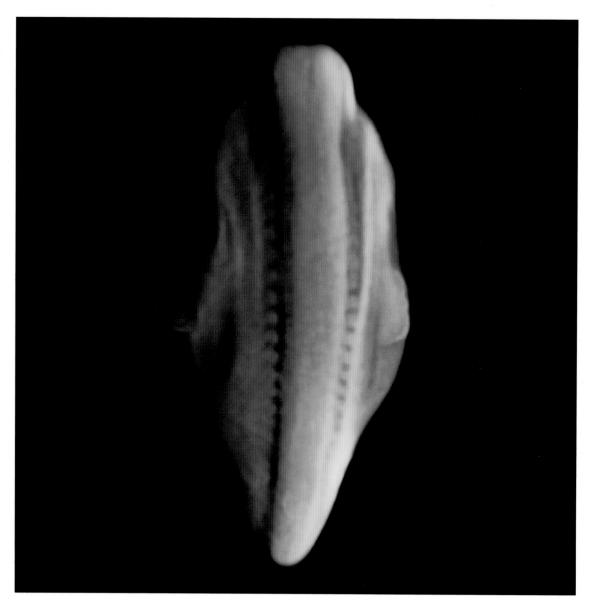

Back view. The baby has a spine and a skeleton.

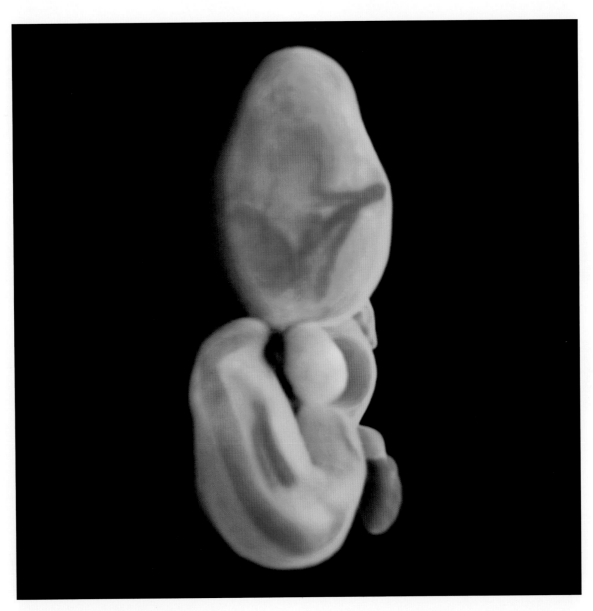

Front view with the umbilical cord.

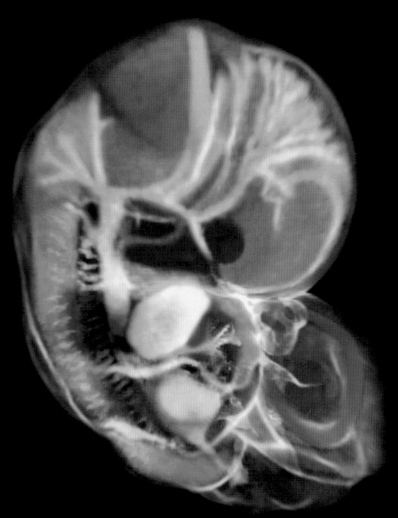

The heart pumps life

The skin is made translucent in these 2 images
so you can see the complexity of the cardiovascular
system (in pink).

44 DAYS

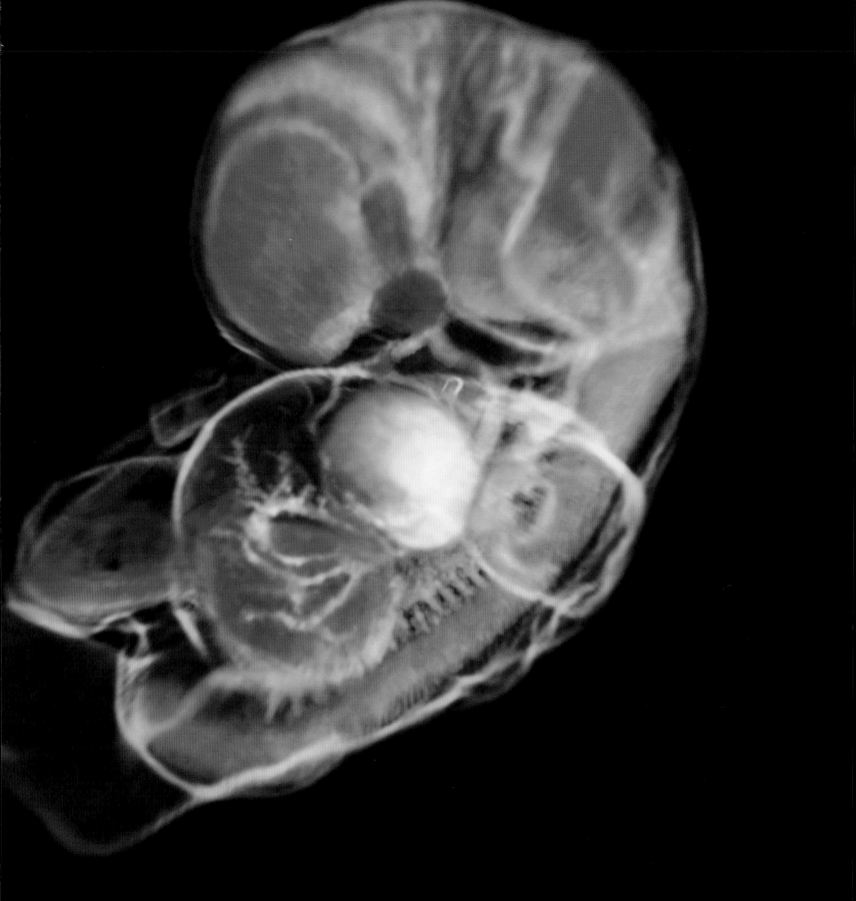

44 DAYS

Umbilical cord

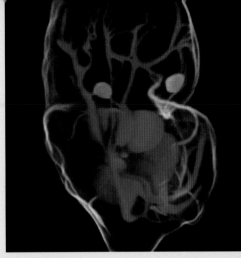

6 WEEKS

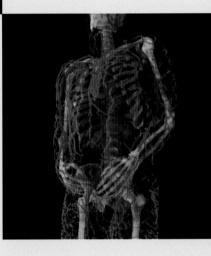

CIRCULATORY SYSTEM

THE HEART'S MAP: The blood transports so many substances and does so many jobs it's hard to name them. Chiefly, though, it moves oxygen and food to all parts of the body while taking away waste chemicals. This process necessarily begins in the second week and by 20 days a system of water-tight tubes branch throughout the embryo. The system grows "ahead" of the baby, proliferating in all directions, like highways on a busy map, stimulating new areas of development. (Understanding how blood vessels are laid down in relation to cellular reproduction has provided some of the most hopeful recent insights regarding cancer.) After birth this network also serves to provide the baby with central heat, distributing warmth evenly through the body, from busy parts like the heart to cooler areas like still muscles.

ADULT
A full-grown human's system is much more intricate than a baby's.

◄ Facing page: The 2 main blood streams are fusing together on the sides of the neural tube at 3 weeks, and a tiny heart starts beating at double the rate of the mom's heartbeat. The umbilical cord provides connection between Mom's and the baby's cardiovascular systems.

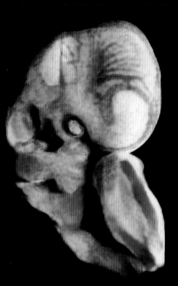
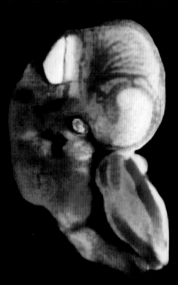

Seeing inside

This new noninvasive visualization
technology can show us, as if in a film
sequence, the inside of a young life
at an extraordinarily early stage. At 44
days, the internal organs are present
but not all fully developed. Ninety-nine
percent of muscles can be identified
at this time, each with its nerve supply.

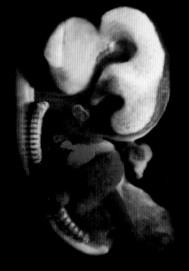

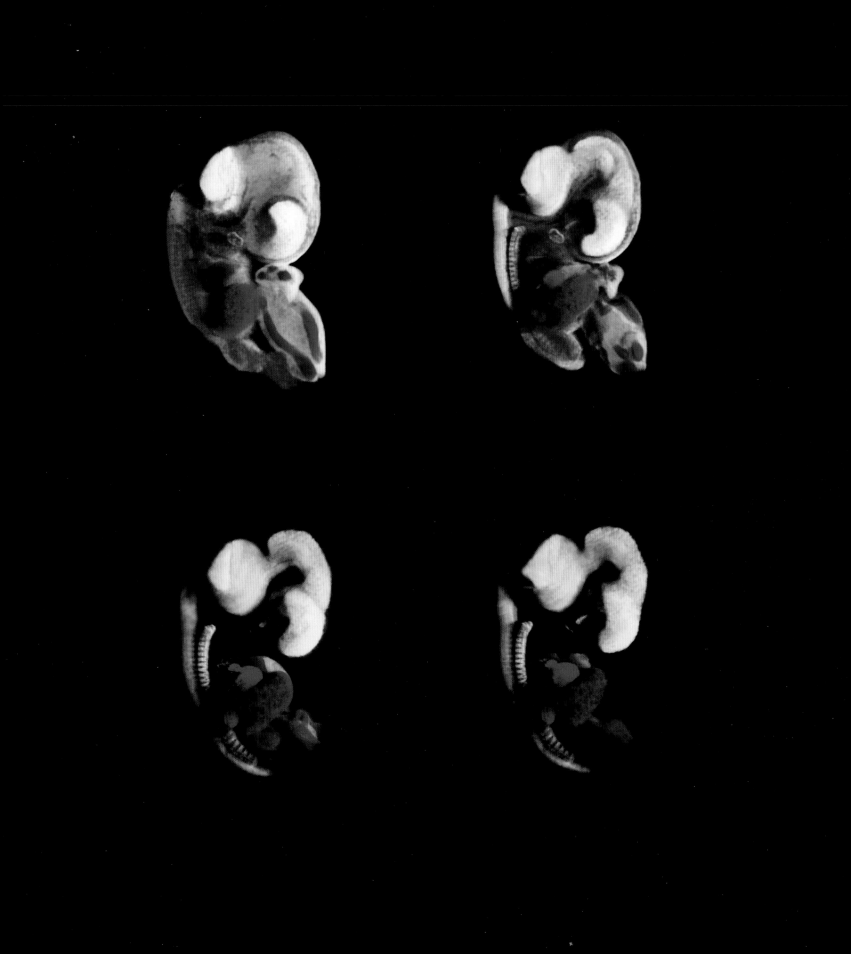

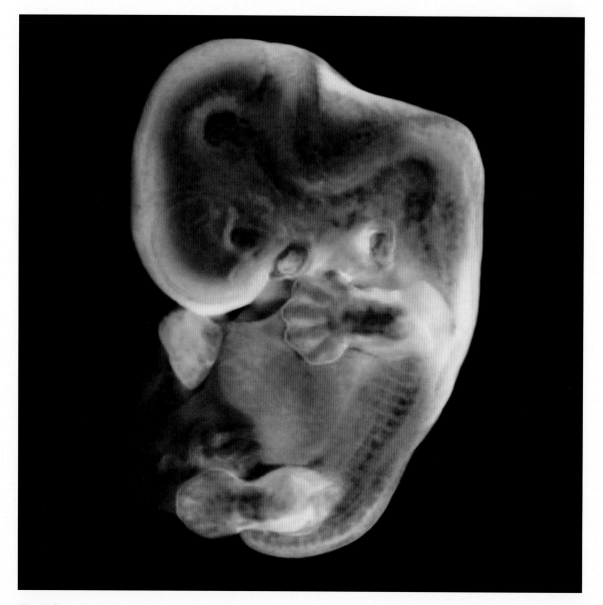

Blood flow feeds the embryo from the mother. At 44 days, for the first time, the baby's kidney reacts to that feeding by producing urine.

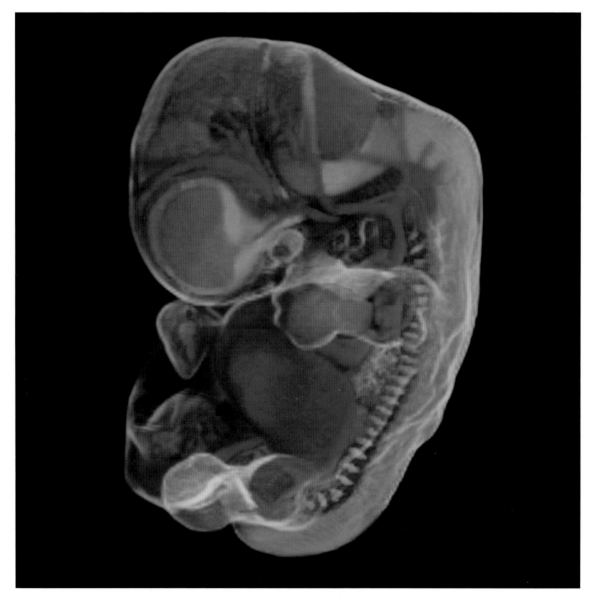

The vertebrae are beginning to develop.

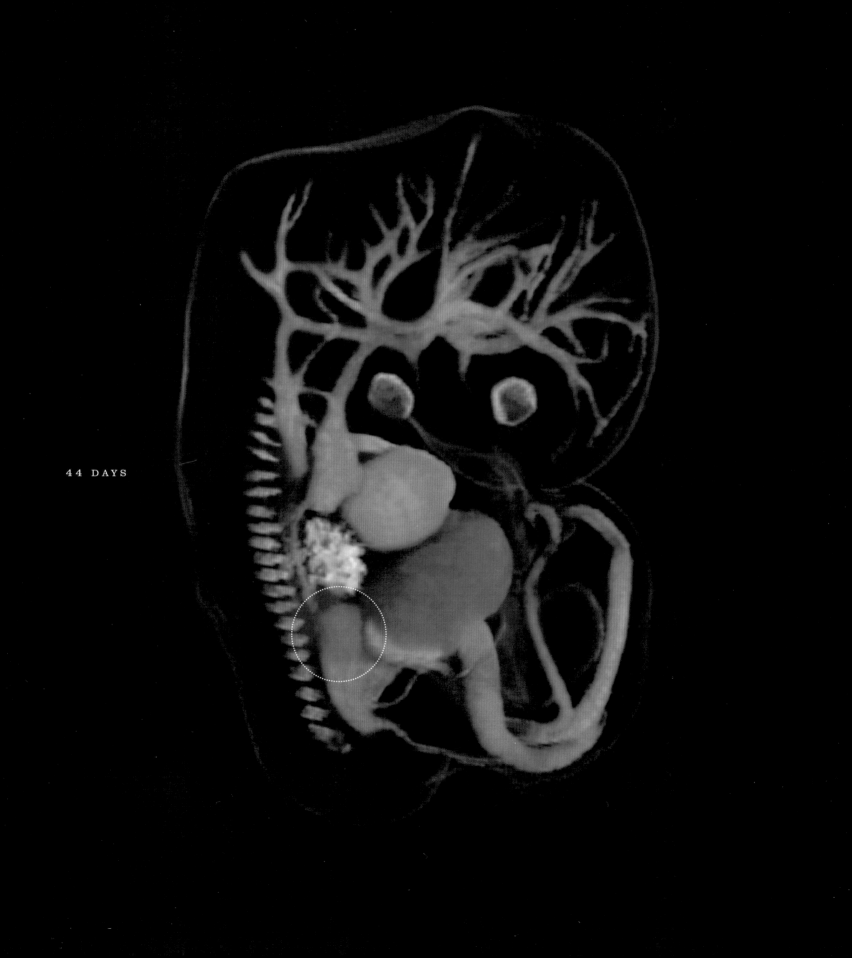

44 DAYS

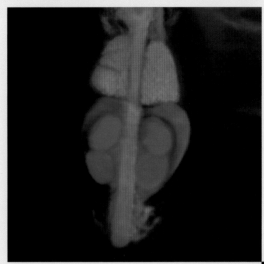

9 WEEKS

ADULT

BUILDING A BABY | 06

THE KIDNEY

CLEANSING: Three successive sets of kidneys eliminate wastes from the blood, the last forming in the third month. These permanent organs must eventually strain 15 gallons of recirculating fluid daily, and it's estimated that they have 2 million microscopic filters to trap particles. Though the kidneys are fully formed in utero, they are among the first organs that must spring to life at birth, when wastes can no longer be passed on to the mother through the placenta. Heroically, many newborn babies will urinate as soon as their umbilical cords are cut.

◄ Facing page: Blood flow feeds the embryo from the mother. The kidneys (in pink) produce urine at week 6 for the first time. Above, see the placement of adult kidneys and note the relative size compared to the abdomen of a 9-week fetus.

Embryo at 48 Days

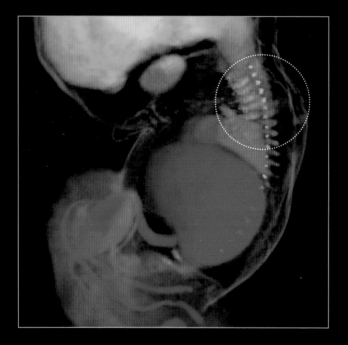

ACTUAL SIZE 16.0 MM

TRANSLUCENCY

The new imaging technology functions as a window into the complexity of the body's systems. In this image particularly, note the skeletal and neural systems.

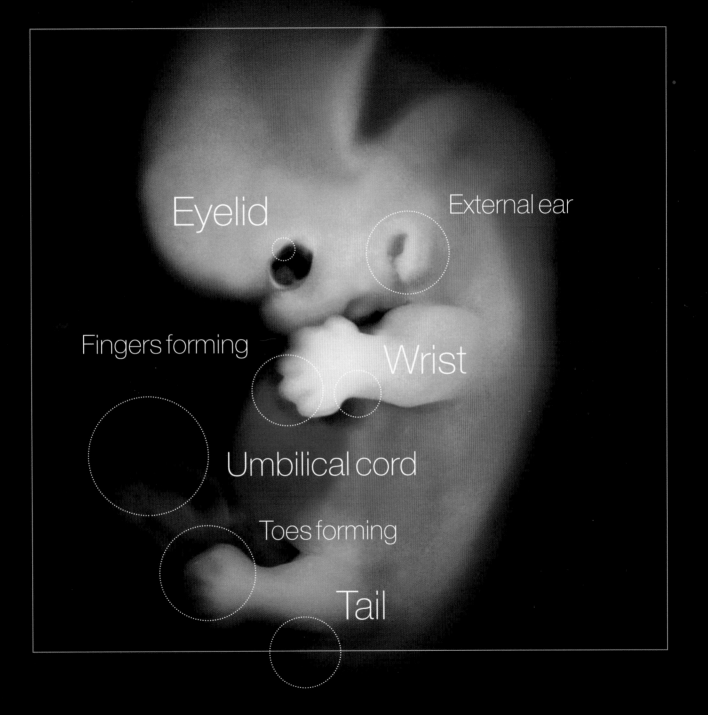

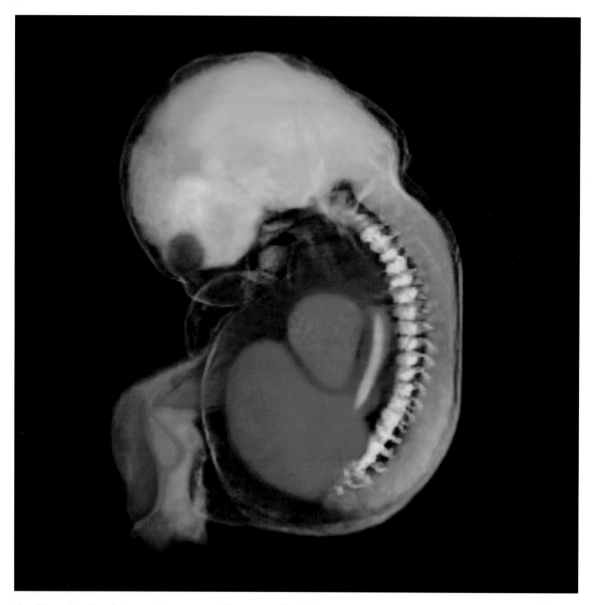

Cartilage develops between bones, providing more flexibility and structural strength to the baby. Also, note the esophagus (the line to the immediate left of the spine).

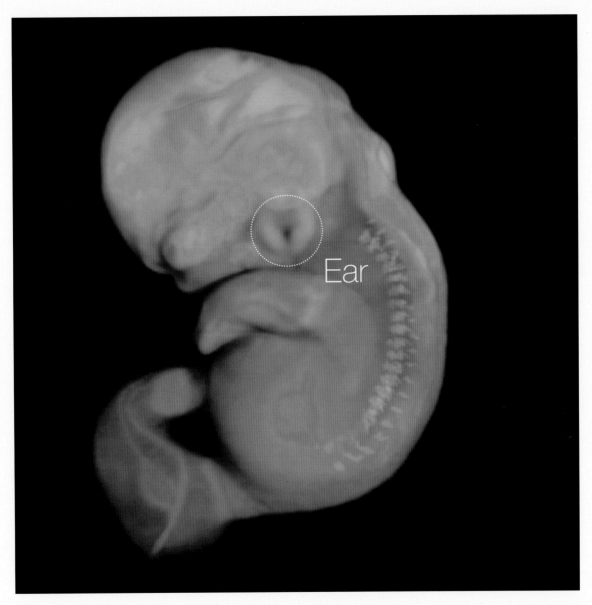

Ear

The ear, now at the base of the skull, will continue to migrate upward.

Floating

The inner ear develops, which will enable the baby
to sense balance and position in the womb.

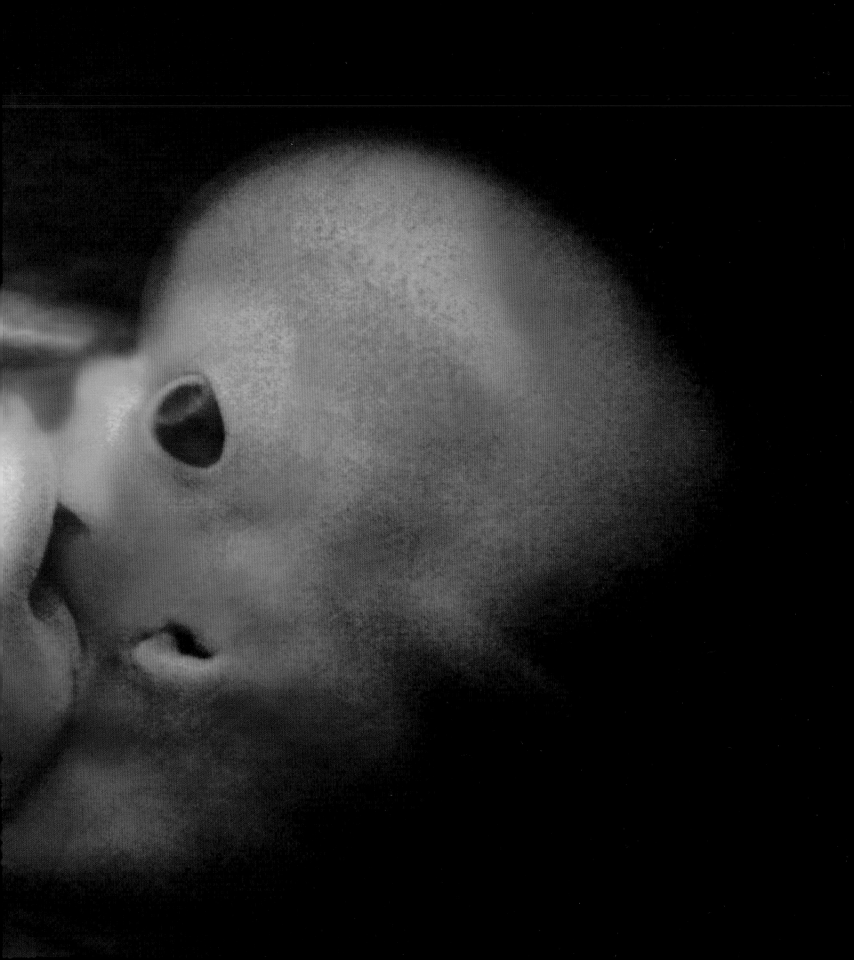

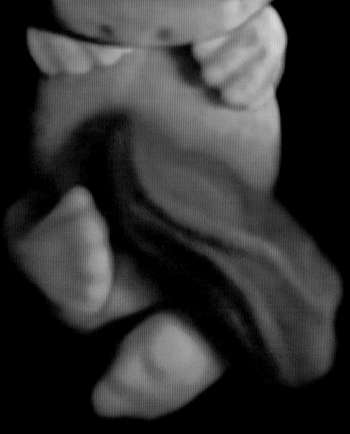

Dependency

The toes have begun to form, but the baby is a long way from walking. You can clearly see the complex flow of nutrients and filtering between mother and child.

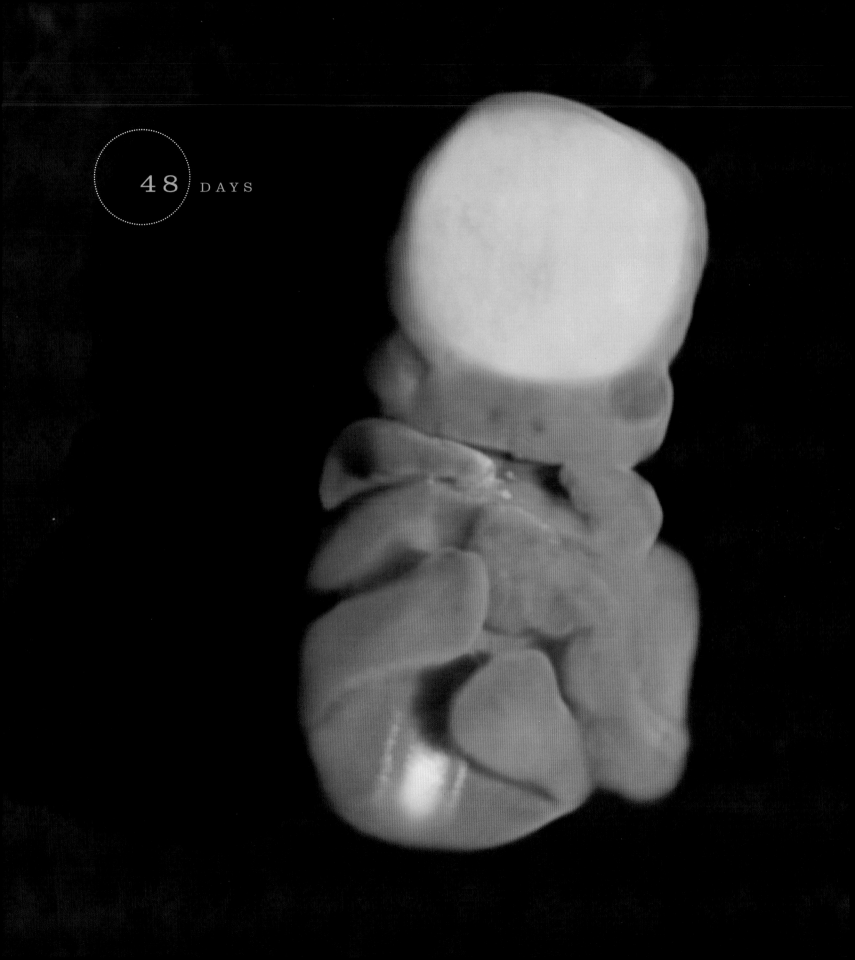

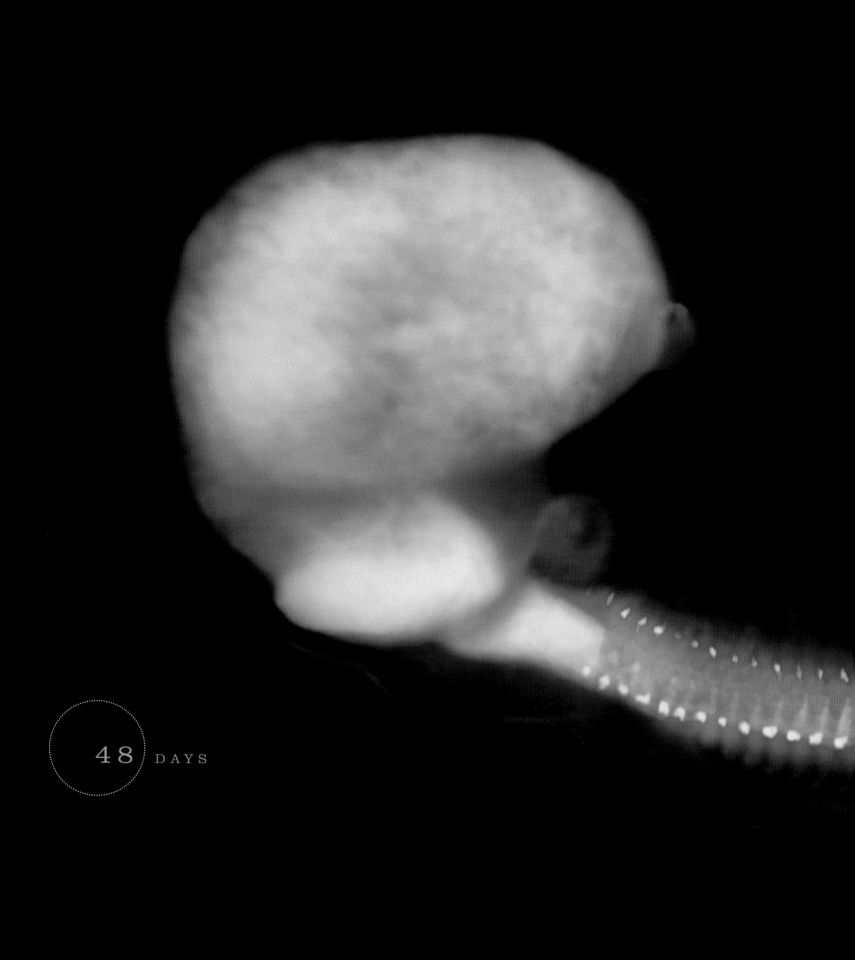

48 DAYS

Electric jewels

Note how the nerve endings use the vertebrae
as a guide wire to feather itself through the rest
of the body's nerve system.

Left: The brain connects with
the muscular system. The baby,
involuntarily, starts moving.

BUILDING A PHYSIQUE

A THOUSAND MARVELS: Upper and lower limb buds develop rapidly during this period, with arm buds leading leg buds by several days at each stage. This seems to anticipate a pattern later on: human infants grasp with their hands long before they can walk. The hands are now far apart, the fingers recognizable yet webbed. The arms are bent at the elbows. Within a few days bone begins to replace cartilage in the shoulders, arms, and legs, as well as the jaws and, to a limited extent, the body skeleton.

As neck muscles begin to form, the head, which has grown large, straightens slightly, and the embryo's original tight C-shape is more elongated. Its posture is that of a night watchman fallen asleep in his chair, head on chest. The main network for supplying blood to the growing brain appears above the eye, near the temple. Permanent kidney tubules replace the embryo's temporary apparatus for filtering blood.

In girls, the ovaries begin to descend.

Embryo at 51 Days

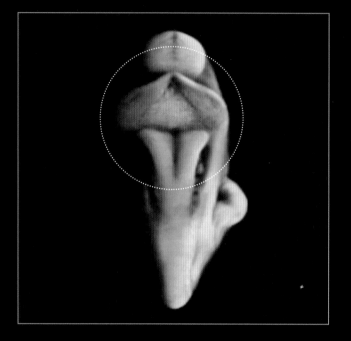

ACTUAL SIZE 18.0 MM

Top view showing the brain tissue closing up over the
fourth ventricle (bright highlight) of the brain.

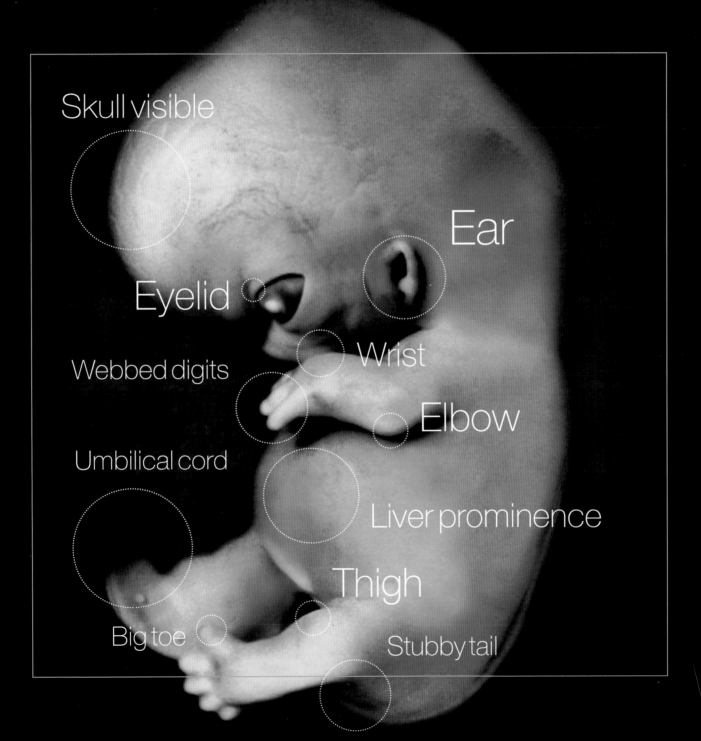

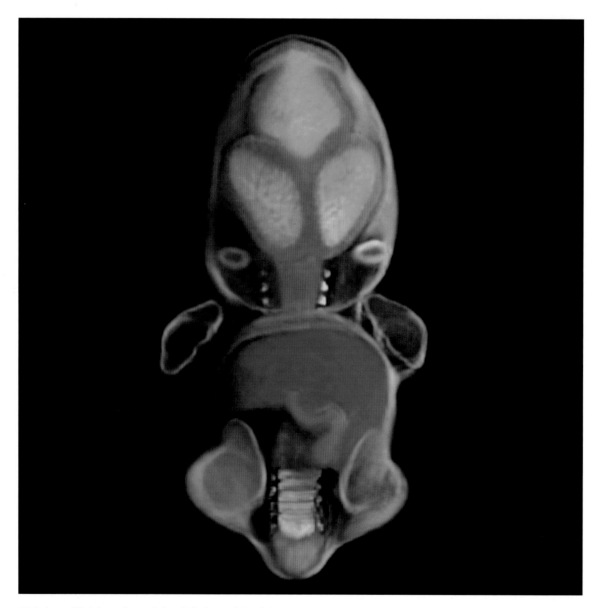

This beautiful translucent frontal view of the fetus shows the bottom of the vertebrae is developing into the lower back bones.

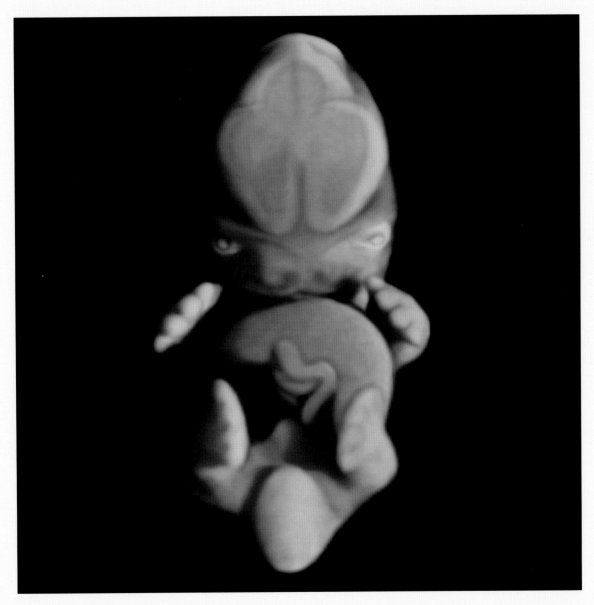

A stubby tail is still present; nasal openings are formed.

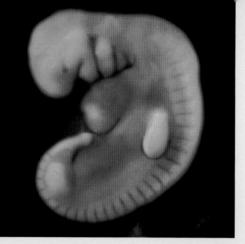

4 WEEKS

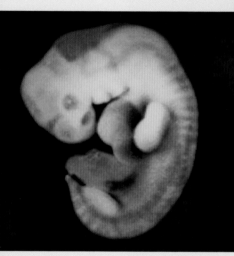

5 WEEKS

At 4 weeks: Arms are flipperlike;
legs begin to appear.
At 5 weeks: Hand begins to develop; legs are now paddle-shaped.
At 7 weeks: Arms bend at elbow,
fingers are webbed, toes appear.

HANDS AND FEET

SO EARLY: The rudiments of fingers appear at about 6 weeks, when the fetus is half an inch long. One week later, all the fingers are present, although not yet fully formed. As the hands spread wide during the next several weeks, fanning like the spikes on the Statue of Liberty's crown, or Lisa Simpson's hairdo, they began to open and close. Fingernails develop; the thumb grows out to oppose the fingers; tiny ridges for gripping (and later identification) form at the tips. By 13 weeks, the hand — which has 27 bones connected by a brocade of ligaments — is structurally complete. One month later, it possesses a firm grip, demonstrating muscular strength, and reflex action coordination. Thumb-sucking may begin around this time, sometimes so avidly that doctors have reported delivering babies with calluses on one thumb.

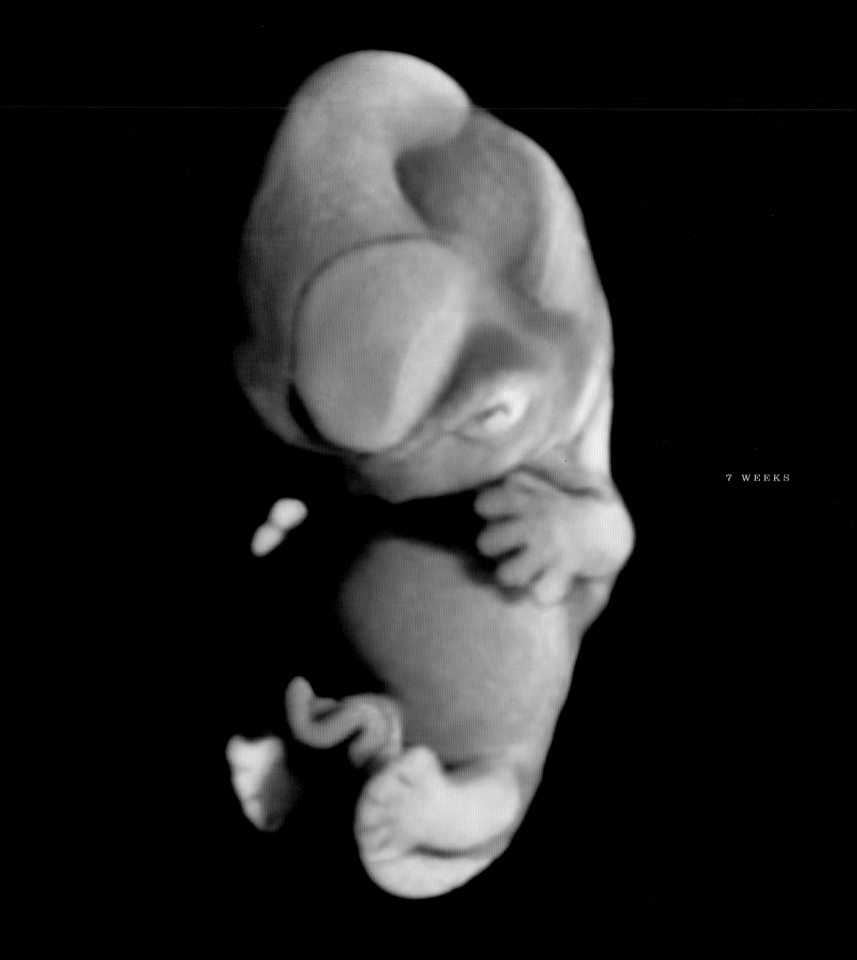

7 WEEKS

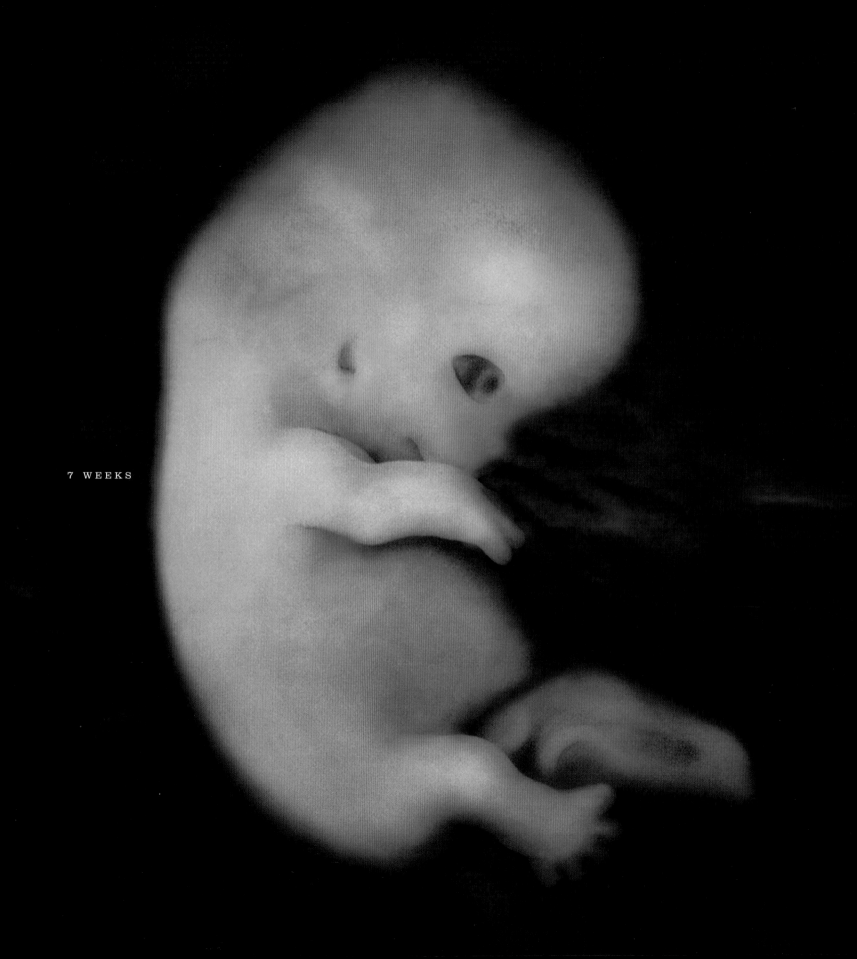

7 WEEKS

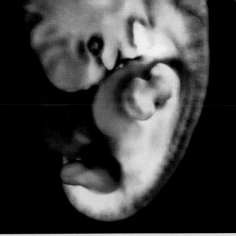

6 WEEKS

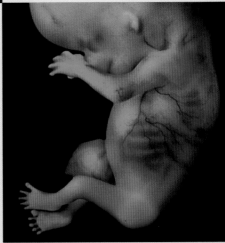

8 WEEKS

HANDS AND FEET

ON TWO FEET: Bones are busy living tissues, and no part of the body has more of them than the foot — 26 in each, lashed together by tough, stringy bands of connective fibers. The baby's feet begin as paddlelike plates at the tips of the leg buds at around 6 weeks, when the embryo is still less than a half-inch long. (Nature builds appendages in the order in which it is going to need them; since the baby will have to grab and hold things before she finds a reason to stand and walk, feet begin to develop about a week later than hands.) By 13 weeks, the toes are articulated, the heel has developed, and the skeletal architecture is fully in place. A miracle of design, the foot is an elastic arch, flattening when you put it down and springing back to a curved shape when you lift it. This helps achieve the uniquely human walking style of swinging our legs in front and behind us as we go around upright — the stride.

At 6 weeks: Fingers begin to develop, as do feet.
At 7 weeks: Fingers and toes are distinct and elongated.
At 8 weeks: Elbows are bent, arm and feet are more defined.

The hand develops much more quickly than the foot.

51 days: Fingers are forming but cannot yet move independently.

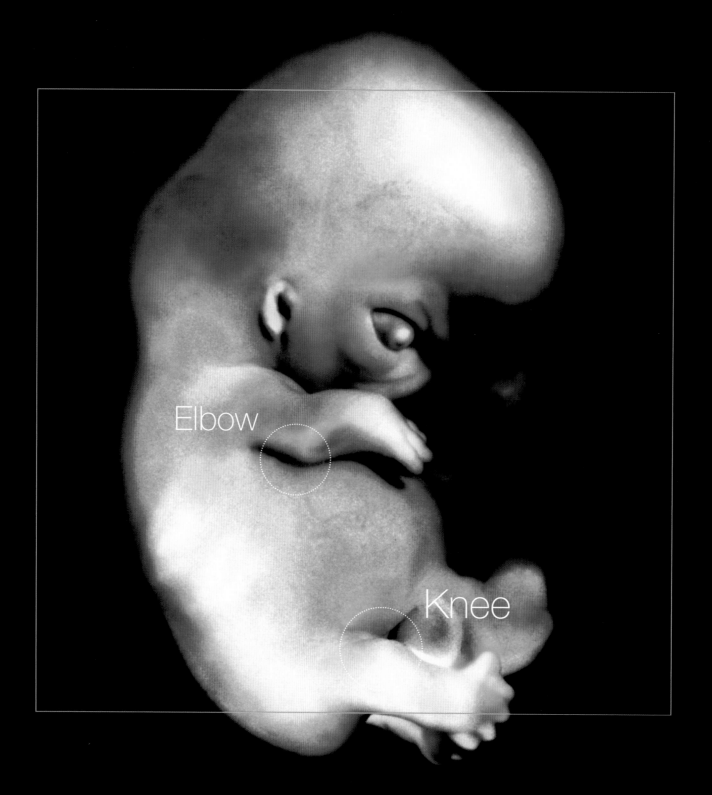

Elbow

Knee

Embryo at 52 Days

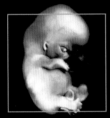

ACTUAL SIZE 26.0 MM

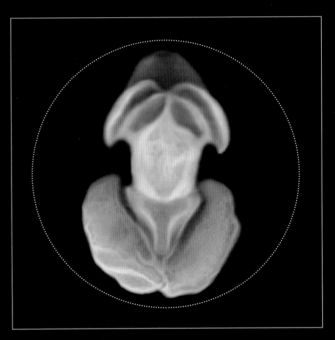

**A VIEW OF THE 52-DAY-OLD BRAIN
VIEWED FROM THE TOP**

At the bottom is the cerebral hemisphere, which will develop into the left and right sides of the brain. The narrow middle section is the midbrain, which houses several constituent parts, including what will become the thalamus. The thalamus approximates a relay station, fielding, interpreting, and directing sensory signals from both the spinal cord and midbrain to the cerebral cortex and sites of the cerebrum. The two crescent-shaped wings will form the cerebellum, the so-called "little brain," which links with other regions of the brain and the spinal cord. It facilitates smooth, precise movement and controls balance and posture as well as playing a role in speech. The tip at the top is the brainstem, which connects the brain to the spinal cord.

Eyes and nose

Fingers are separated, eyes strongly pigmented.
Very clear, open nostrils.

52 DAYS

The skull

The plates of the head are clearly forming now.
The eyes are still on the side of the head but move
to the front as the head enlarges.

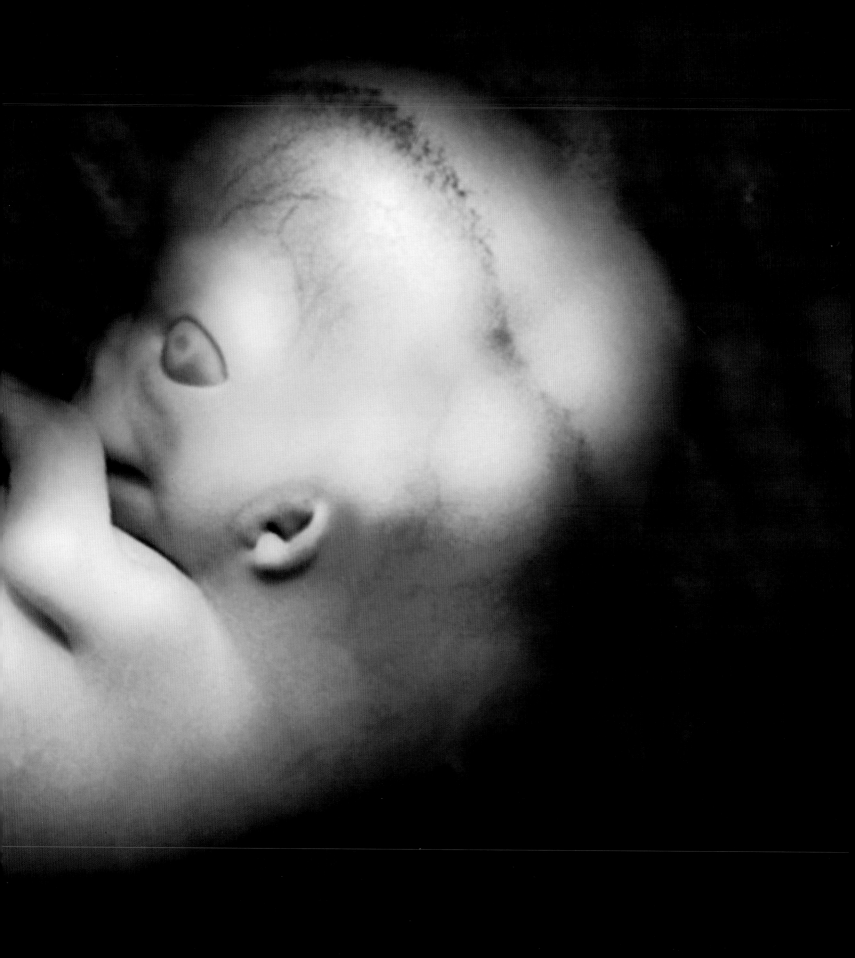

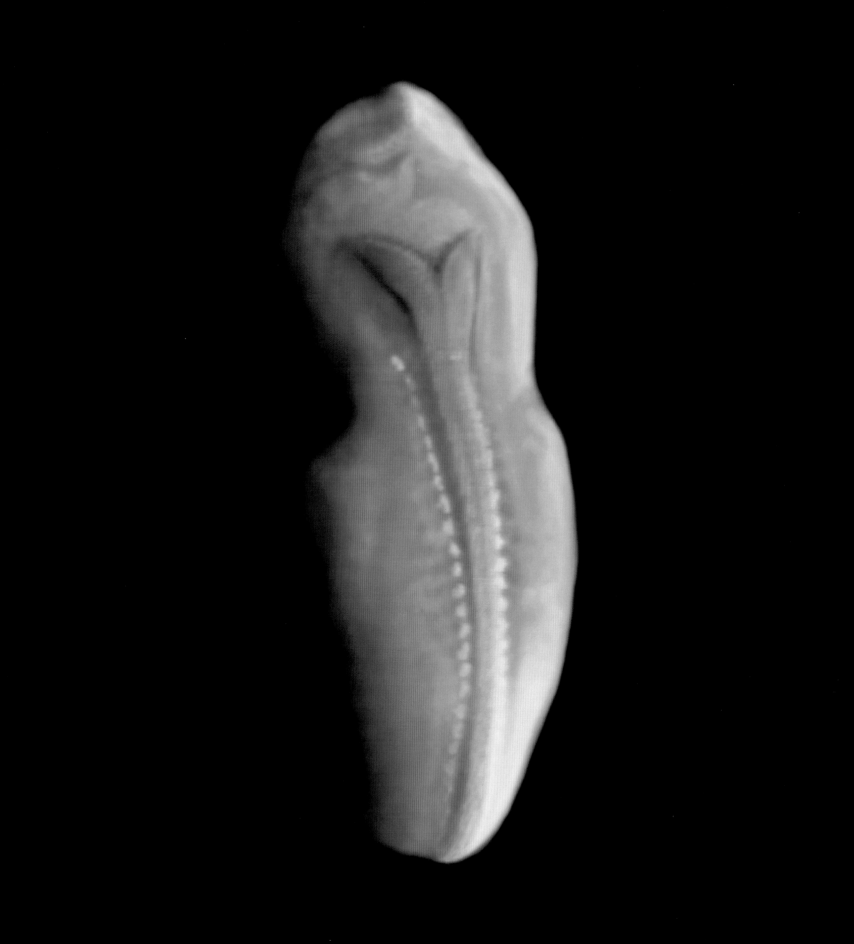

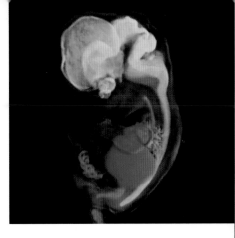

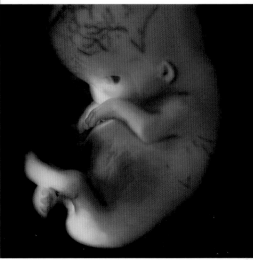

TRANSITIONS

AS THE EYES MOVE from the side of the head to the front, folds of skin grow rapidly over the corneas to protect them. Skin also now grows over all remaining facial grooves, smoothing the appearance. With the external ear and nose almost fully formed, the face looks unmistakably human, although the jaws remain undeveloped.

Partitioning of the heart, which began in the fifth week, continues as the intricate moving architecture of chambers, channels, shunts, and bypasses is laid down. Meanwhile, the major blood vessels of the body assume their final plan.

With the head held more upright, the arms clasped over the bulging midsection, and the legs extended and crossed, the embryo assumes the familiar posture known as the fetal position.

54 DAYS

Embryo at 54 Days

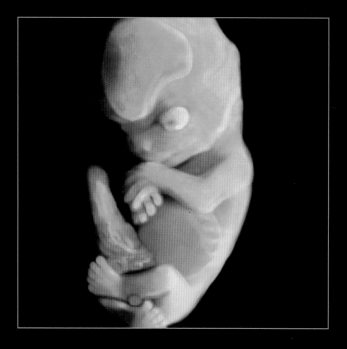

ACTUAL SIZE 26.0 MM

THE BRAIN

Proportionately, the brain and central nervous system take up far less of the baby's body mass. The embryonic brain has yet to develop the iconic folds and wrinkles we recognize in adult brains.

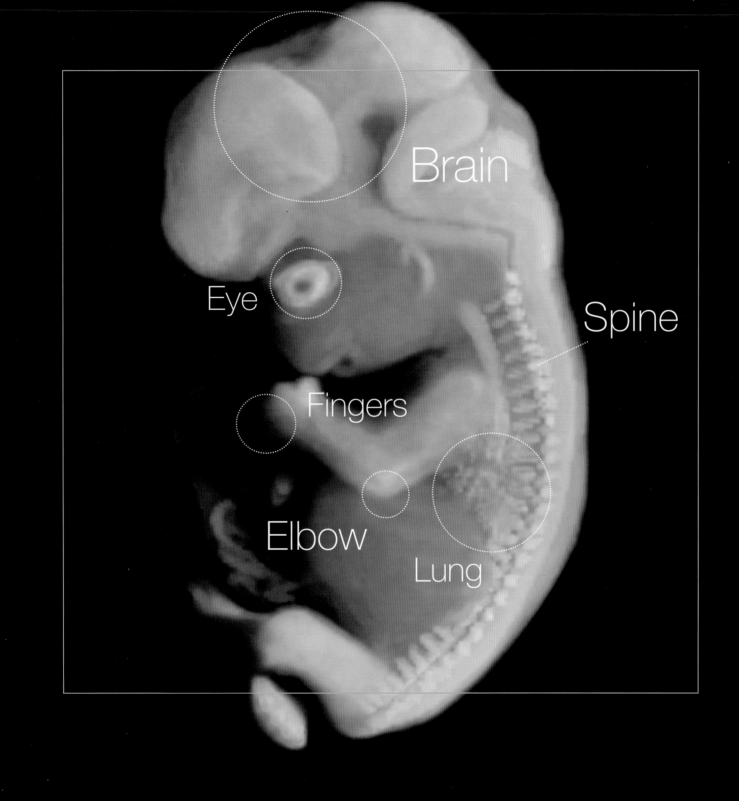

Bones

The skull, knee, shoulder, and elbow definition
show how the developing bone (cartilage) is becoming
more human in appearance.

54 DAYS

Lips

Among many other momentous changes, you can now see that the upper lip is formed.

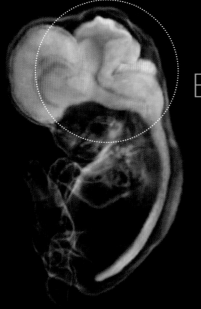
Brain

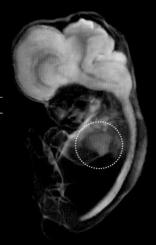
Heart

Organs in development

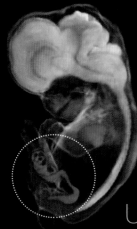
Umbilical cord

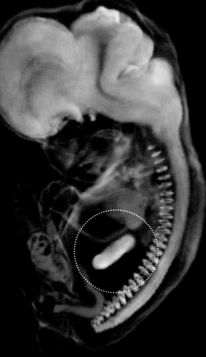
Stomach

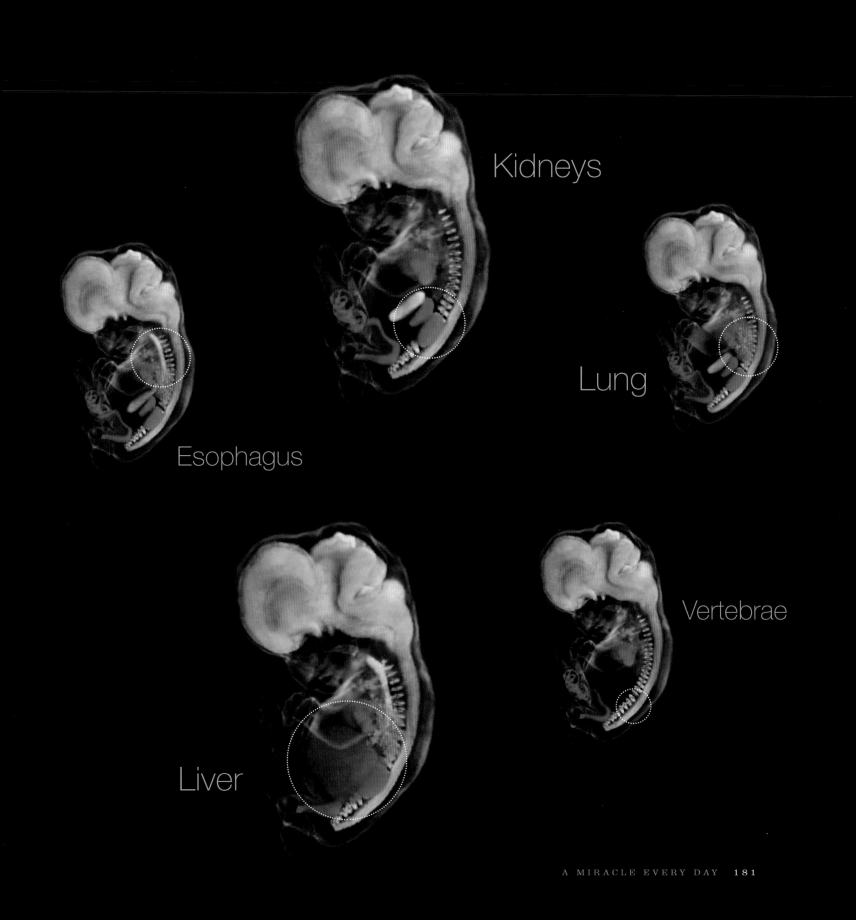

Kidneys

Lung

Esophagus

Vertebrae

Liver

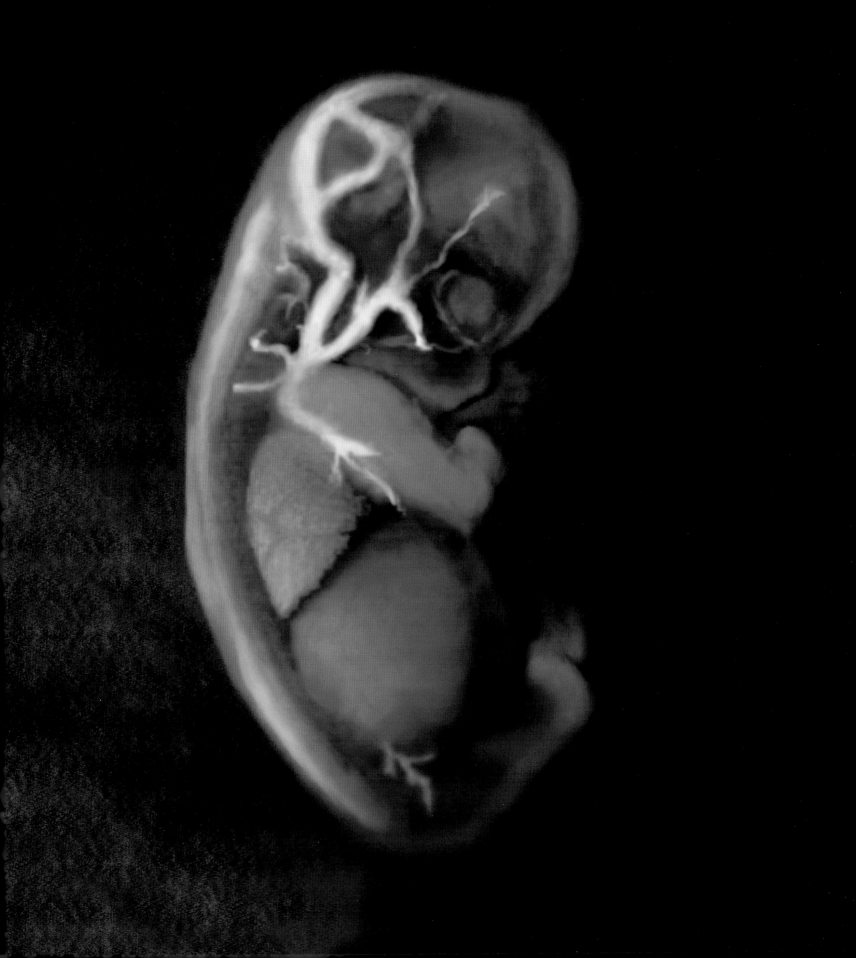

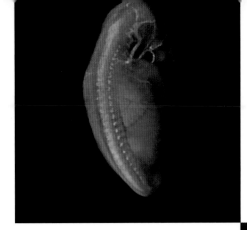

A MOMENTOUS
PASSAGE

EVERYTHING IN PLACE: Measuring one-and-one-fourth inches from crown to rump and weighing about one-thirtieth of an ounce, the embryo is now all but fully formed. All body systems are in place and elaborated. Architecturally, the organism is more or less whole. In the mouth, taste buds develop. The diaphragm emerges, separating the heart and lungs from the intestines. Though the energy output is about one-fifth that of an adult, the heart is functionally complete. The lungs possess lobes and are filigreed with many-branched tubules.

A great passage has been made. All the body's myriad parts — cells, tissues, organs, systems — have been differentiated. Now they will begin the complex process of becoming interconnected and integrated, giving rise to a functioning whole — an active human fetus.

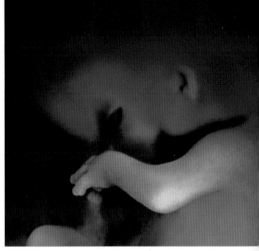

Head is rounded, ears are more fully developed, the spine is beautifully articulated.

◄ Facing page: Note how well developed the lungs are at 56 days.

56 DAYS

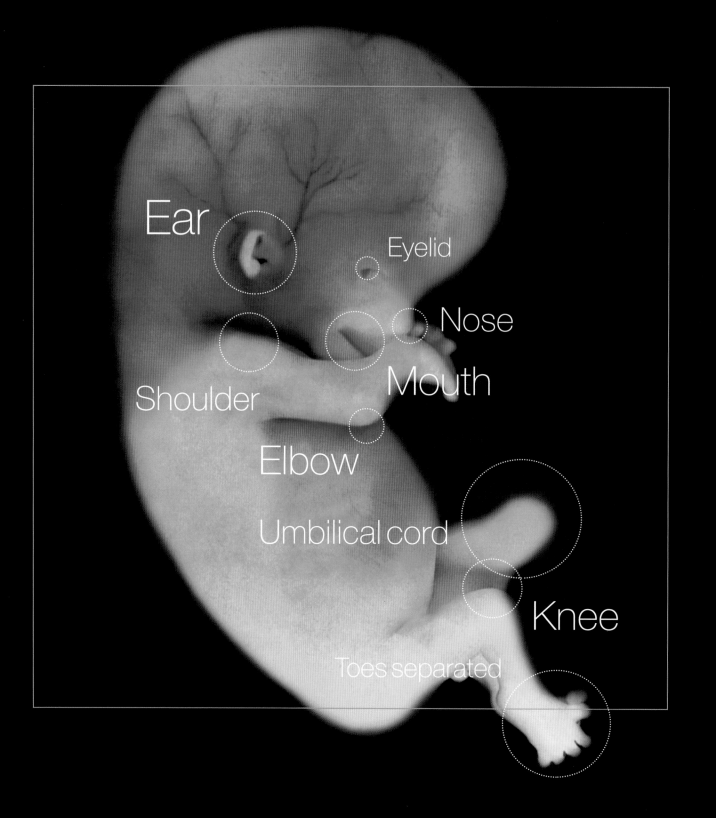

Ear

Eyelid

Nose

Mouth

Shoulder

Elbow

Umbilical cord

Knee

Toes separated

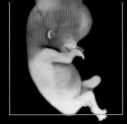

ACTUAL SIZE 30.0 MM

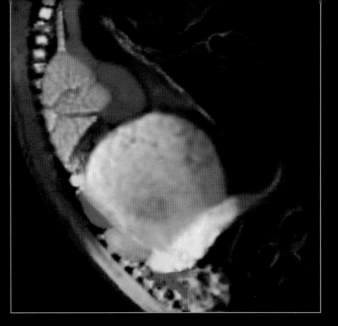

THE ORGANS

All organs in place and defined. The intestines move
out of the umbilical cord and integrate into the embryo's
body cavity.

At home in the womb

Baby in the placenta.

56 DAYS

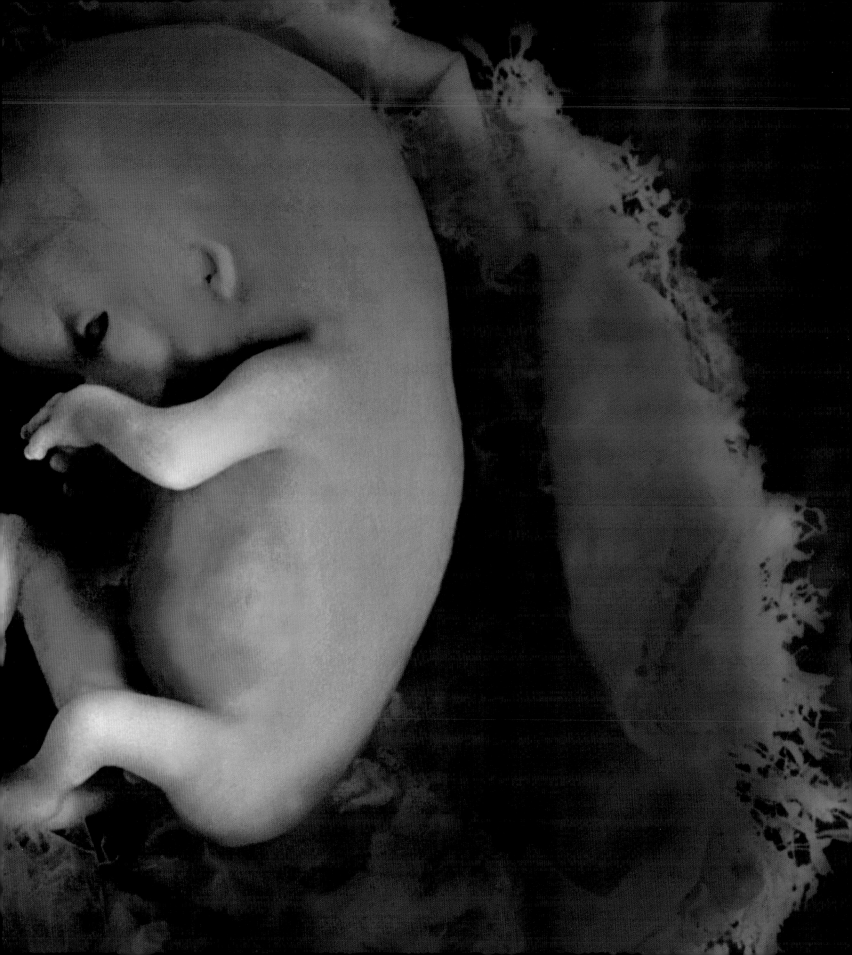

THE LIVER

THE SUBSTITUTE: Nature above all else is resourceful. It makes a living by devising exquisitely suited anatomical structures. Yet it's also a virtuoso at making do with what's available. The liver is the body's largest and most complex glandular organ, a chemical megaplex where digested proteins and fats are converted into dozens of life-sustaining substances. However, since before birth there's nothing to digest, the embryonic liver does short-term duty as a proxy, a temp, churning out red blood cells until corpuscle-producing marrow can be formed.

ADULT

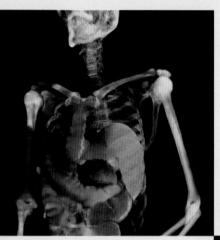

14 WEEKS

> Facing page: The liver (in pink) is taking care of the production of red and white blood cells during the pregnancy. This function decreases during the last 2 months. The weight of the liver is 10 percent of the body weight around 10 weeks and still 5 percent at birth.

Left: Note how large an adult liver is. Compare to liver at 14 weeks.

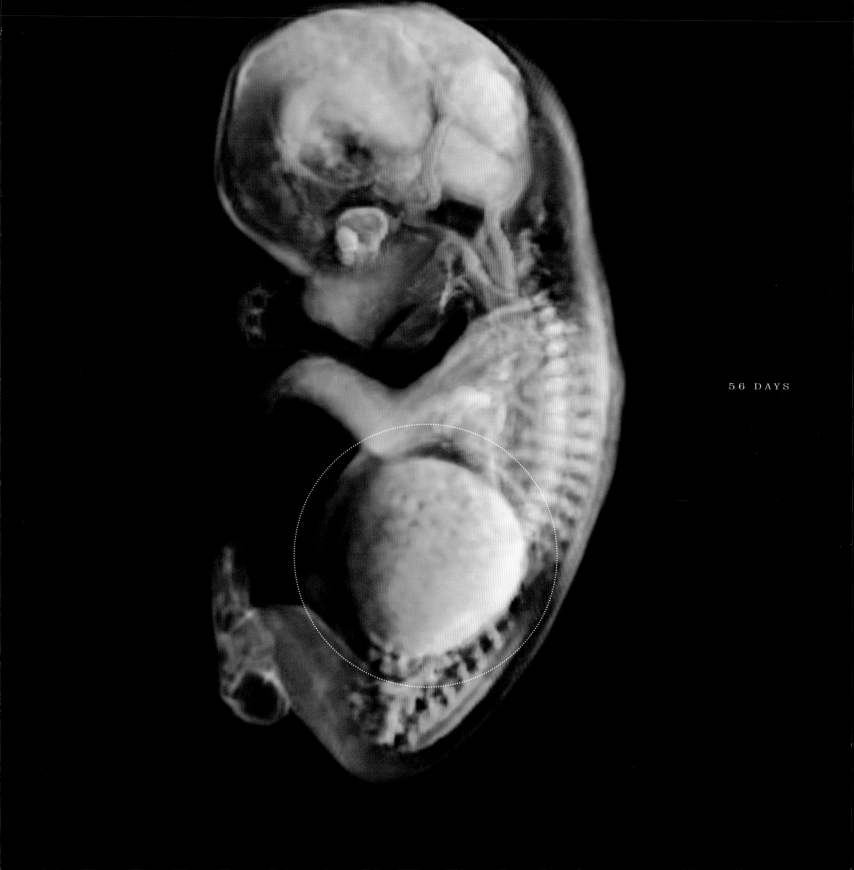

56 DAYS

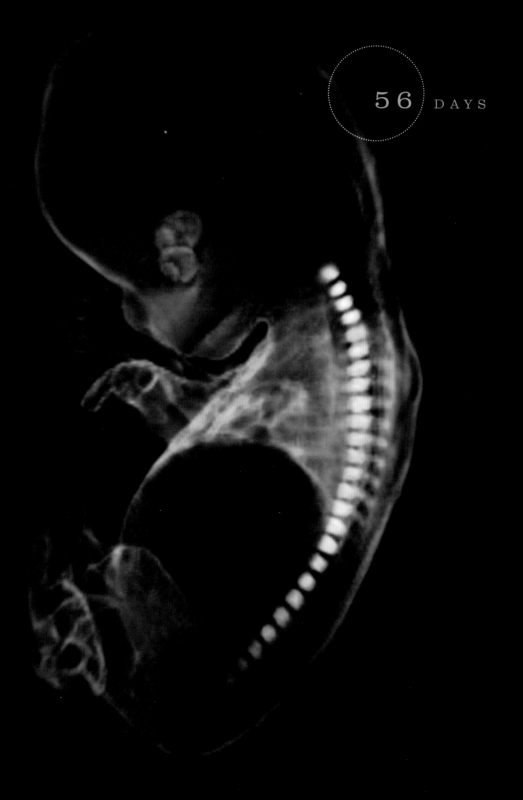

The spine

The little backbone provides strength
and structure to the embryo.

Messages along
the spine

The nervous system and nerve endings.

Nerve endings

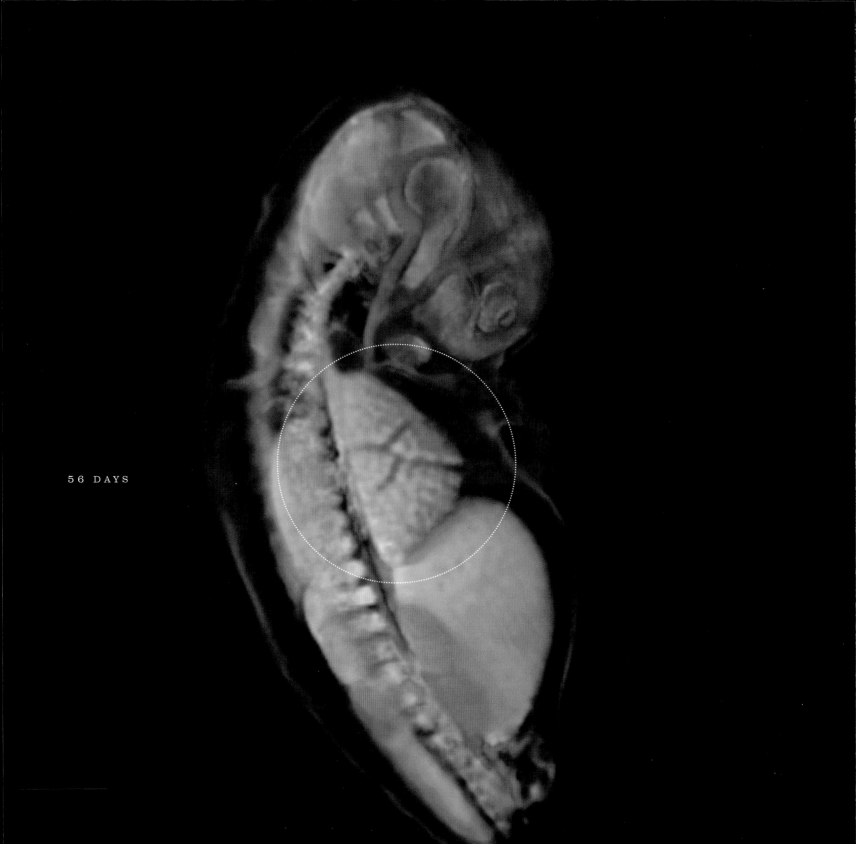

56 DAYS

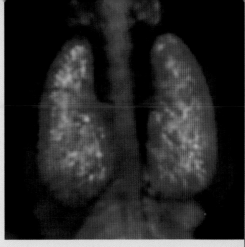

8 WEEKS

At 8 weeks (above), the tubes in the lung that will feed oxygen to the blood for the rest of our lives are now clearly formed.

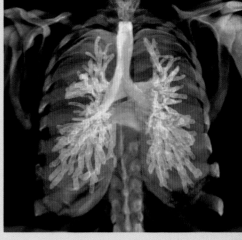

ADULT

◄ Facing page: Lung of an 8-week fetus (yellow). A human's left and right lung are not identical. Here we can see the left lung, which already has developed its 3 lobes. (The right lung has only 2.)

> BUILDING A BABY | 09

THE LUNGS

The respiratory system first arises during the fourth week as an inconspicuous groove at the rear of the facial region. From there, it grows downward, branching 23 times on both sides of the body, forming so called "long buds." At 4 weeks the first buds appear and the lung starts developing. The tube for swallowing food (esophagus) separates from the breathing (laryngotracheal) tube. At 6 weeks secondary branches appear inside of the lung. These buds continue to branch after birth, eventually connecting 300 million tiny air bubbles — alveoli. If all your alveoli were spread out flat, they would cover a tennis court.

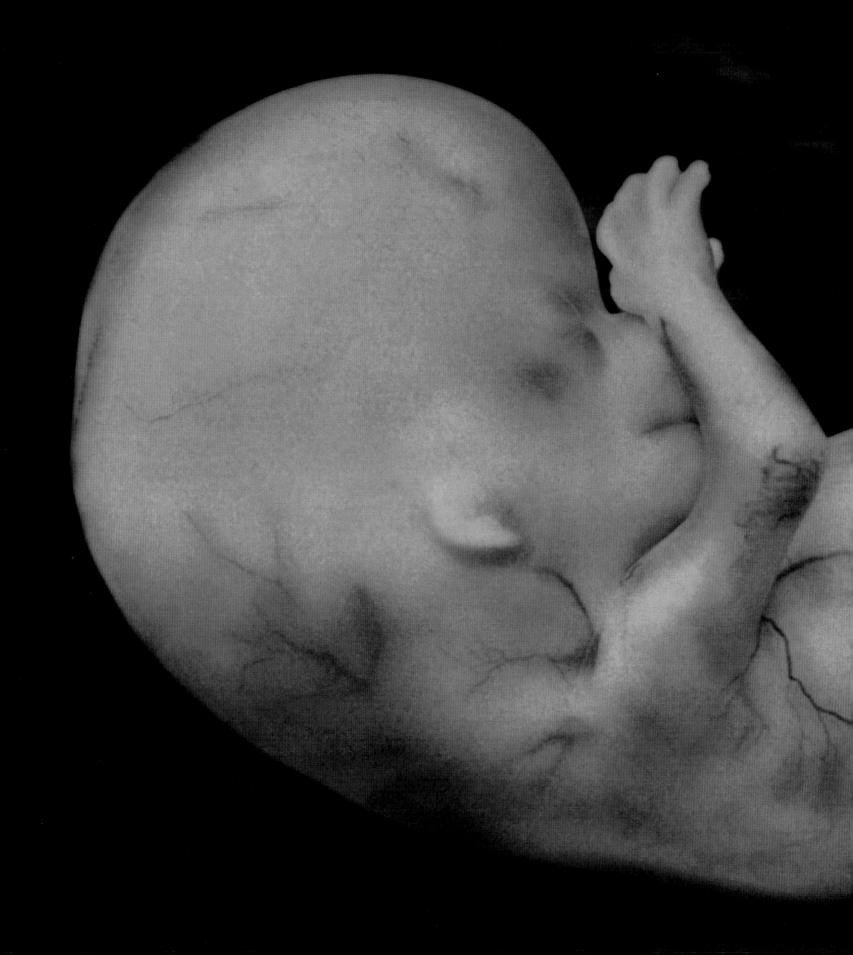

56 DAYS

The next stage in the journey

From here on in, development is more about
growing than body systems forming.

Growing

SENSING THE OUTSIDE WORLD

MOVEMENT

GETTING STRONGER

Intestines have migrated into
fetal cavities of the abdomen from
the umbilical cord.

04

Activity

Despite its wondrous growth and development, its physical achievement, the fetus so far has shown little or no discernible independent movement. It has differentiated into parts—organs, muscles, systems—each with a capacity to perform a specific function. The marvel now is how those parts begin to reunify; how they mold themselves into a form that anticipates, to the last detail, the day the newborn will leave the protective comfort of its mother and begin to fend for itself.

Gradually, at about 8 weeks, the fetus begins to stir, as if from a paralyzing sleep. "Now when the brain signals," embryologist Robert Rugh and obstetrician Landrum Shettles note, "the muscles respond and the fetus begins to kick, turn its feet, and curl its toes." Though it measures little more than an inch long, its arms bend at the elbows and its fingers gather in a fist. Its eyes are sealed shut, months away from first sight. Yet as Rugh and Shettles, pioneering observers, noted, when researchers began in the 1970s to snap more and more portraits inside the womb, it frowns, squints, furrows its brow, purses its lips, and opens its mouth.

These activities lack "purpose" — there is no specific correlation between stimulus and response. But as the fetus stirs, it jars our understanding: *In order for even the healthiest to become an individual, able to survive outside its mother, there remains an astonishing amount of growth, preparation, and refinement still to master.*

And so while the 2-month-old fetus may be more or less whole, it will require another 7 months inside its mother before it can fully face the world — 7 months in which its weight will multiply up to 1,000 times. It is as if all the instruments of an orchestra have been created. Now, in the third month, rehearsals begin for a lifelong symphony, and the players must learn to perform. In unison.

The pregnant mother, for her part, can't feel what's going on inside her uterus, but her body, too, is adapting to the advancing needs of the fetus while paving the way for others still to come. If she has had morning sickness, she can expect it to last another month or more, as pregnancy hormones continue to incite wholesale bodily changes. Meanwhile, her

blood volume starts to build as more nutrition and oxygen are needed by the fetus, and as her body stocks up against an anticipated blood loss during delivery. Her heart gets stronger and beats faster, in order to push this added red cell mass through the placenta. She may have gained a few pounds, probably as a result of her surging hormones, but nothing in comparison to the 25 to 40 pounds she is likely to put on during the next 7 months as she and her baby ready themselves to part.

As for the expectant father, everyone has stopped caring about him and his slighted ego.

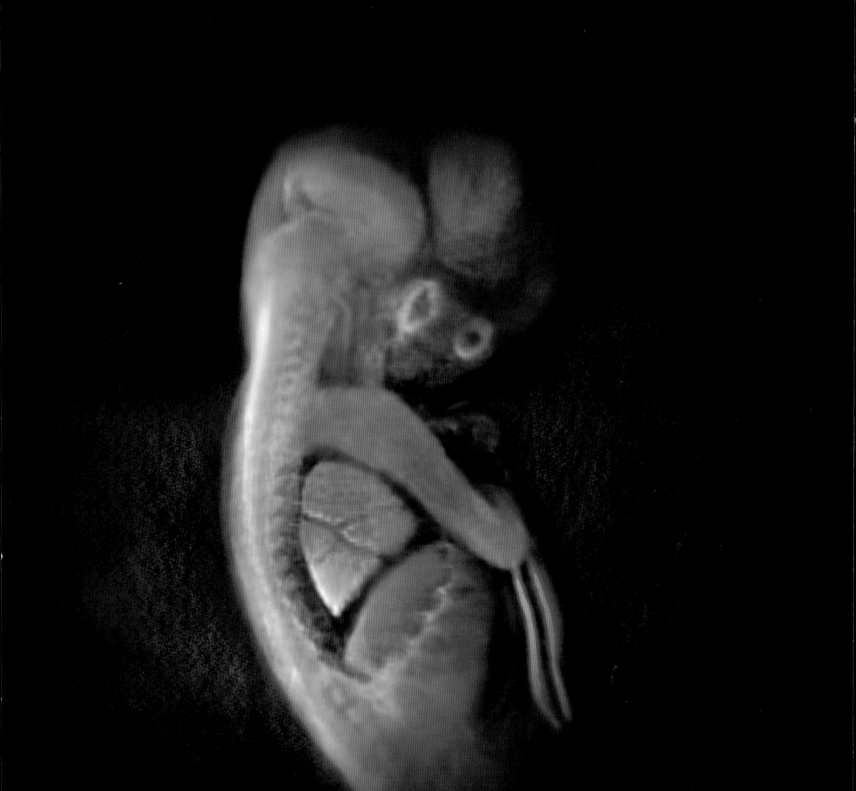

FEELING A PULSE

STIMULI: The most noticeable changes at about 2 months are in stature, which in the past week has become markedly more human: head more erect, back straighter, abdomen tucked. The muscles and skeleton are forming rapidly, and as the body uncurls, the intestines, which previously had grown too quickly to be enclosed and had protruded outside, start returning into the body cavity from the vicinity of the umbilical cord. It may be possible with a stethoscope to feel the fetal pulse.

Though the fetus measures one-and-a-half inches from crown to rump and weighs about one-seventh of an ounce, fingernails and toenails have begun to grow. The skin thickens, acquires layers, becomes less transparent; hair follicles develop below the surface. In the tube-shaped neural canal in which the spinal cord is forming, nerve cells lengthen and intertwine. The male fetus is recognizable by the budding of an external penis, although the female genitalia do not begin to show for another week.

As the muscular and nervous systems continue to integrate, the fetus begins to respond more specifically to stimulation. Prodded, the eyelids and the palms of the hands now close.

◄ Facing page: Large, more rounded head, visible neck, and lowly positioned ears are characteristics of a fetus between 9 and 12 weeks.

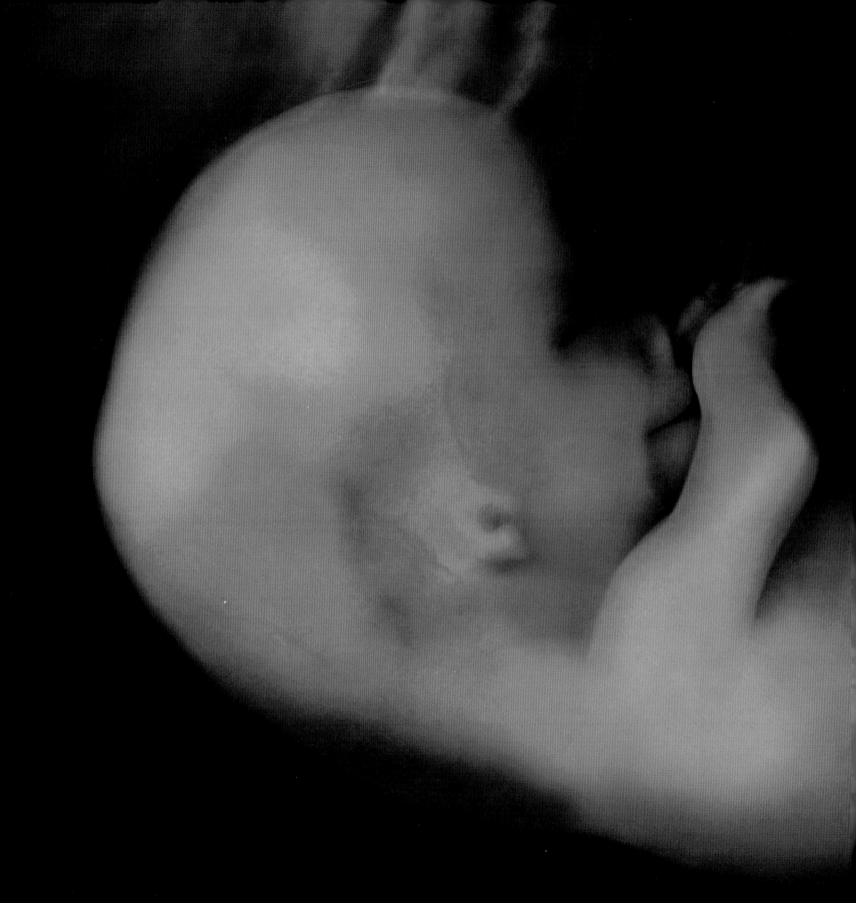

Protecting heart and eyes

The rib cage begins to close and eyelids are closed from 9 weeks.

TEETH AND OTHER WONDERS

COMING INTO PROPORTION: Several more organs mature to the point of taking up their permanent roles — coming on line, so to speak. The thyroid gland, the master switch for running the body's complex chemistry system (metabolism), clicks on, as does the pancreas, which makes digestive enzymes. Even though there is no food to digest, the gallbladder begins to secrete bile into the digestive tract, where successive muscles are forming in preparation for the day when the newborn must process food on its own. The branching of lung tissue is nearing completion. Straddling the heart, the thymus, seat of the immune system, is infiltrated by cells that ultimately will become the body's chief sentinels — T-Cells.

As bones form in many places, even fingers and toes, cells in their hollows start churning out blood cells, which first were manufactured outside the body in the yolk sac and later in the liver and spleen. The number of connections between muscles and nerves has tripled since week 9. The face, which appears more human as the eyes move forward and the ears up on the head, is sensitive to touch, as are other regions of the body. The first permanent tooth buds form.

The fetus has grown considerably, measuring a little more than 2 inches in "sitting position" and weighing about a quarter of an ounce. Now that the lower regions of the body are growing faster in comparison to the head and the intestines have withdrawn fully inside, its proportions begin to look more infantlike.

Right: During this period (9-12 weeks) the fetus doubles its length, and the intestines begin to reach their final location.

Intestines

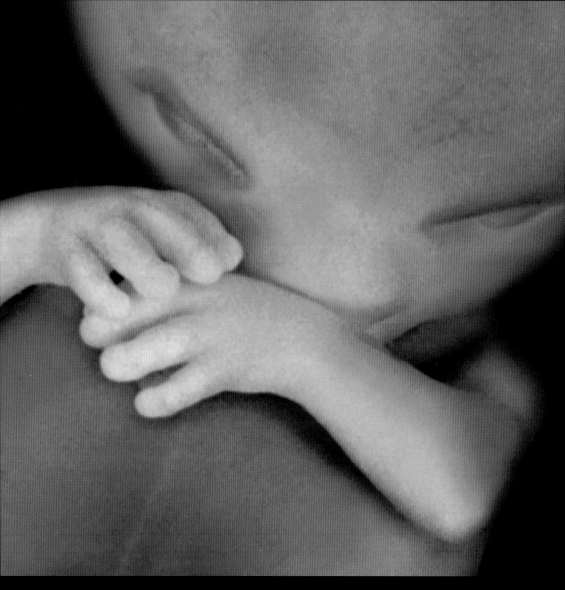

Fingernails begin to grow from nail beds at week 10.

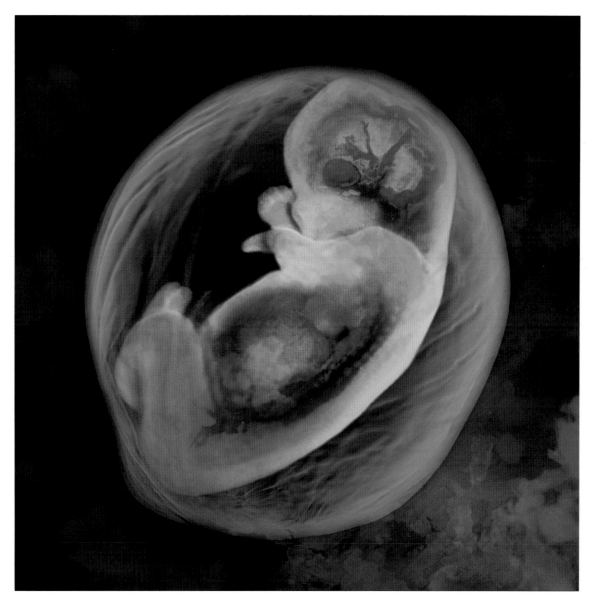

Liver starts to secrete bile, a thick, brown-green liquid containing bile salts, bile pigments, cholesterol, and inorganic salts. The bile is stored in the gall bladder. Development of thyroid, pancreas, and gall bladder is complete. Pancreas starts to produce insulin. Volume of amniotic fluid at week 11 is approximately 70 grams.

INDEPENDENCE

GETTING READY: Fetal development can been seen as a race to build and refine systems to enable the newborn to live eventually outside the womb—before the child is expelled. The defining goal is independence. Already functions that were first assigned to the mother, or carried on through the placenta, are being brought "in-house."

Trachea, lungs, stomach, liver, pancreas, and intestines are all actively developing into their final and functional shapes. Though the mother's circulatory system remains chiefly responsible for transporting nutrients, protecting against disease, and scavenging waste, the fetal blood supply now begins to help out. Glands in the pancreas start producing insulin to break down sugar and starch. The lining of the intestines form microscopic folds, increasing the amount of surface area available for absorbing nutrients into the blood. Eventually each fold will be lined with 3,000 cells, the intestines will grow to 32 feet, and 20 million glands will secrete enough digestive enzymes for the body to absorb, say, a 24-ounce steak.

Corresponding preparations are underway in the mouth. The jaws, like other bony structures, have begun to harden—ossify—and are strung with "tooth germs," infinitesimal buds that will grow into milk and baby teeth. (In order for the teeth and bones to be properly formed, women especially need additional calcium during this period, and during the rest of pregnancy.) The vocal cords of the larynx—the voice box—also begin to form, even though the fetus, which still measures about two-and-a-half inches and weighs perhaps a third of an ounce, remains a half-year or more away from having an air supply to enable the vocal cords to vibrate and make sounds.

➤ Facing page: Skin is very sensitive, reflexes are working, the head is still one-third of the length of the body.

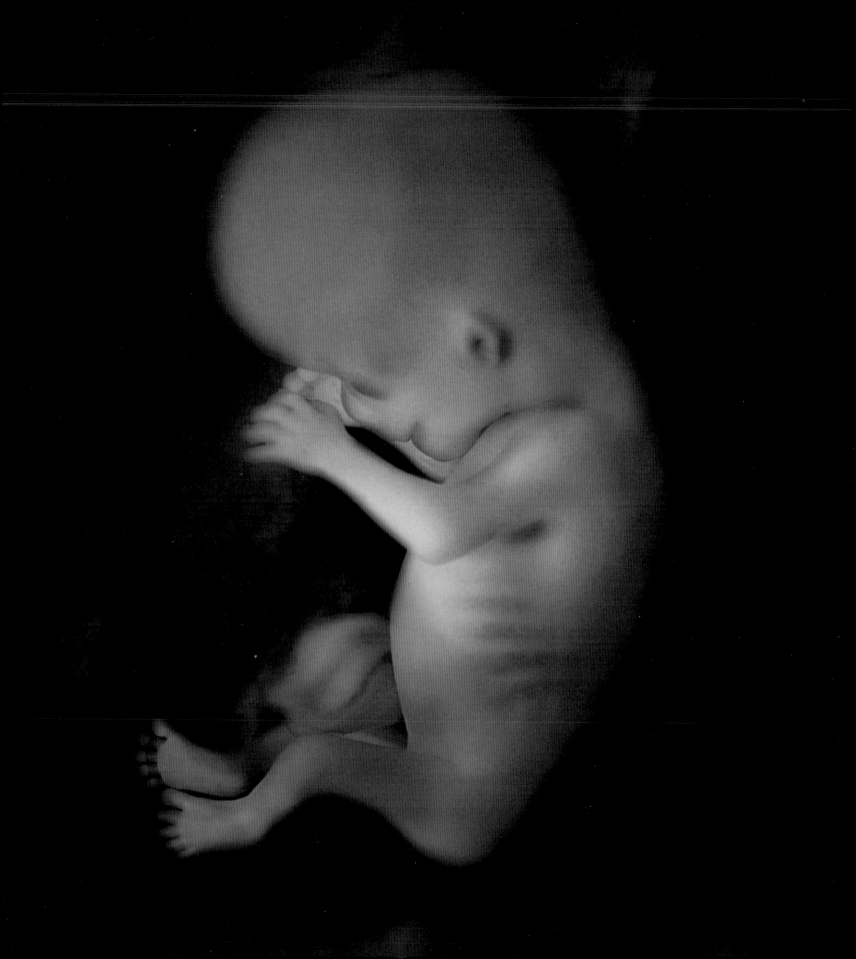

Bigger every day

By week 12 the weight of the fetus increases
to about a tenth of an ounce.

12 WEEKS

INDIVIDUALITY

A PERSONALITY: By the end of the first trimester, the fetus has developed all of its major systems. No new organs remain to be formed and those that already exist will spend 6 more months becoming what they need to be for the newborn to survive. The fetus is virtually complete in form and actively functioning, although still far from being able to succeed in the world on its own.

Like other primates, it now has an opposable thumb—an enormous achievement, as Rugh and Shettles have written, "because it allows us to pick up objects, hold a pencil or paintbrush, or manipulate a dial on a spaceship panelboard." Its motions are more purposeful, anticipating critical future activities, like breathing and eating. The eyes have moved closer together on both sides of the nose, which has developed a bridge.

Perhaps most remarkable is the emergence of a rudimentary individuality. The fetus is not yet conscious, but its uniqueness begins to assert itself. As Rugh and Shettles wrote: "Fetuses of the same age begin to show individual variations, based in all probability on behavioral patterns inherited from the parents." Different fetuses now make different facial expressions.

Left: Sucking muscles of mouth fill out cheeks, tooth buds continue to develop, and salivary glands begin to function.

Mind

Structurally the brain is developed; from
this point on its mass increases greatly.

Heart

The heart has remarkably achieved its basic
adult shape, structure, and blood flow.

Strength

Eyes and ears become more forward, neck
is straighter, stronger at 14 weeks. The head
starts turning.

Movement

Fetus begins to move around, though the
mother can not yet sense these movements.

13 WEEKS

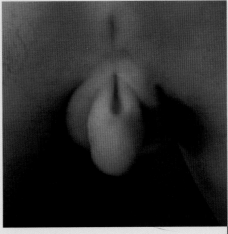

If the urethral groove (the opening in the middle) closes up then a boy will be born.

If the urethral grove stays open, a girl is about to arrive.

◄ Facing page: The genitalia of the fetus develop from week 4 and become external from week 8.

THE INDIFFERENT PENIS

GENDER: As scientific nomenclature, the term seems almost comical—a sexual disorder among twentyish slackers who wear black and affect a studied ennui, or else men over 70. But in biology, indifference doesn't mean boredom: it means unmediated sameness. Up until 5 weeks after fertilization, the embryonic sex organs in boys and girls are identical. Genetically, sex has been assigned, but under a microscope the genital regions are indistinguishable—they haven't yet "differentiated" into male and female. When they do, in the third month, the initial protrusion that develops is shaped and molded in such a way that it can become either a penis or a clitoris, with characteristics familiar to both.

3 MONTHS

Boy or girl?

Though the indifferent penis is visible, the sex of the baby
is not clear from external appearance until week 12.

PREPARING FOR THE FUTURE

MUSCLES, BONES, BRAIN, GENES: The beginning of the second trimester signals a change in fetal priorities, from rapid development to rapid growth. During the fourth month the fetus almost doubles in length (from 3½ to 6 inches, in sitting position), while its weight quadruples (from 1 to 4 ounces). As when the blastocyst first attached itself to the uterine wall and gorged on its mother's blood, launching the process of differentiation and development, growing suddenly much larger takes on a special urgency.

So does the need for an improved posture. Up to now the fetus has remained curled, its oversized head slumped forward on its chest. Foreshadowing an upright existence, the body suddenly begins to elongate as its head becomes relatively smaller. To hold the head erect, neck and back muscles develop as the bones of the spine, rib cage, and shoulders knit together and harden. The legs lengthen rapidly, although not the torso and arms.

Though the head becomes less prominent, its primacy expands. The surface of the forebrain—the cerebrum, seat of the intellect, motor control, and memory—fills with pleats and wrinkles, increasing its usable area. The underlying regions of the brain that control emotion, hunger, sexual drive, balance, and other basic impulses emerge but remain less developed.

Meawhile, the body continues to prepare not only for its own future but for the future of its genes. In the female, the "accessory" sexual organs—fallopian tubes—are fashioned from the ducts left behind by the degenerating first set of primitive kidneys. The uterus and vagina attach, although the vagina is packed temporarily with a mass of cells.

The skin is thin, loose, and wrinkled, yet on the tips of the fingers and toes distinctive ridges and whorls appear. No two individuals, even identical twins, share the exact same patterns.

Limbs are well developed.

Long-limbed

The arms and legs are beautifully elongated at 16 weeks.

4 MONTHS

THE BABY KICKS

ANNOUNCING ITSELF: Fetal movements — kicks, turns, even hiccups — are now pronounced enough to be felt by the expectant mother, who may be able to identify an elbow, or a head, nudging against her abdominal wall. At 8 inches in length from crown to rump and weighing about half a pound, the fetus has grown to about the size of a squirrel. Though most organs are well advanced, the baby remains unprepared to live independently, as certain essential features — such as the emergence of sweat glands to help regulate body temperature — remain to be developed.

Facing page: Fingertips and toes develop the unique swirls and creases of fingerprints and toe prints.

Most bones are now hard. A faint ridge of eyelashes appears. In both sexes pale pink nipples develop over the mammary glands, which are equipped with milk ducts. It may seem as if, in boys, the ability to produce milk is wasted, but nature seems midway during pregnancy to be actively sorting out what ultimately will — and will not — be of use to the child. More taste buds, for instance, have already been produced than will be needed at birth, and the fetus already possesses a firm handgrip. Scientists speculate these may be evolutionary remnants, left over from the development patterns of our prehistoric ancestors, who, for instance, may have needed taste buds on the roof and walls of the mouth and even in the throat to detect poisons in foraged vegetation, or a ferocious grip for clinging to the fur of a mother who traveled high in forest canopies.

As the fetal heart daily pumps 144 liters of blood through the body, its sounds become sufficiently loud to be heard by dads-to-be, siblings-to-be, and others.

The umbilical cord

Circulation is completely functional. The umbilical cord
system continues to grow and thicken as blood travels with
considerable force through the body to nurture the fetus.

5 MONTHS

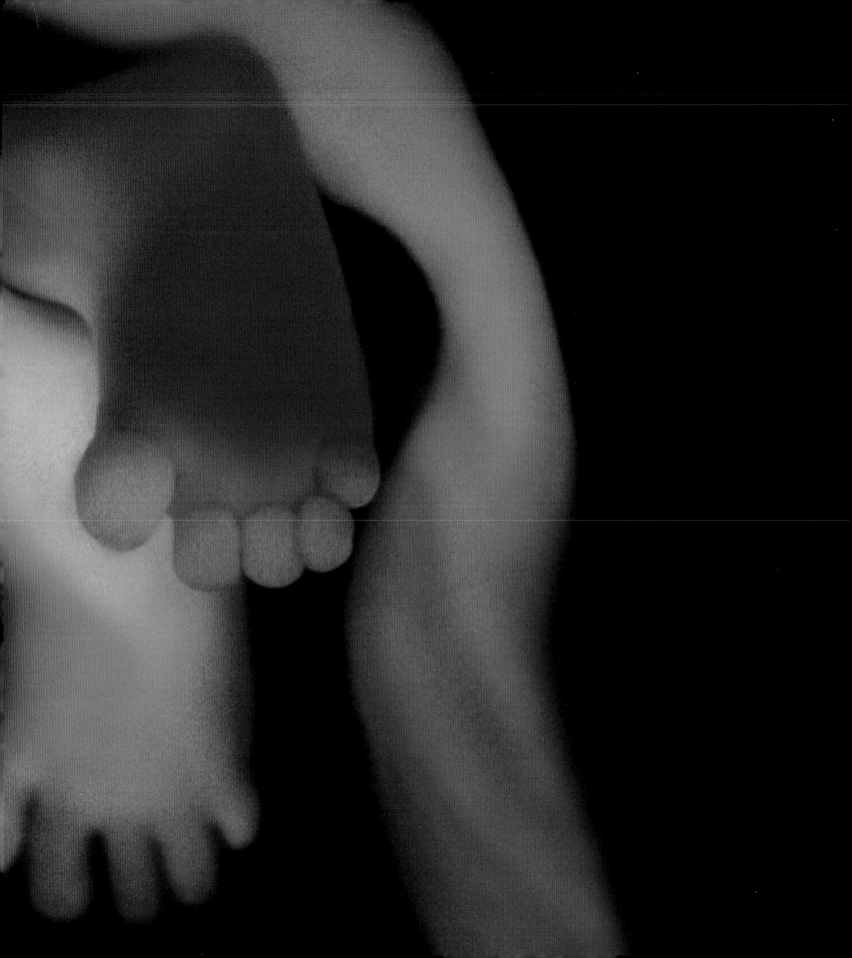

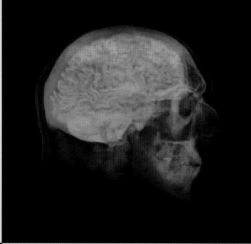

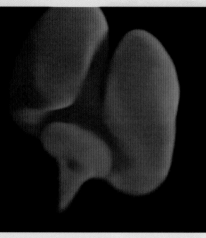

Look at the wrinkles in an adult brain (above right) compared to the smooth fetal brain (above).

➤ Facing page: Side and bottom view of a 20-week-old fetus brain. Though its shape is already closer to its final one, it is still smooth and has no definition yet. Its development is ongoing years after birth.

THE BRAIN

TINY MINDS: "The brain," Emily Dickinson wrote, "is just the weight of God." By the seventh week, nerve cells in the brain have begun to touch via projecting molecules. Some start to form primitive pathways. The rate of production is stunning: 100,000 new cells a minute. At birth, the brain will consist of about 10 million densely entwined nerve cells—some 100 billion cellular units in all. Though single cells measure only one one-hundredth of a millimeter across, each is connected to thousands of others by fingerlike branches along which it can send and relay signals—perhaps 10 trillion connections in all. Connectivity, not mass, accounts for the brain's exceptional abilities. "It has been estimated," Rugh and Shettles wrote, "that all 2 billion of the specific nerve cells which make any individual educable are located in the outer covering of his brain, its cortex, and that these 2 billion cells could be stored in a thimble."

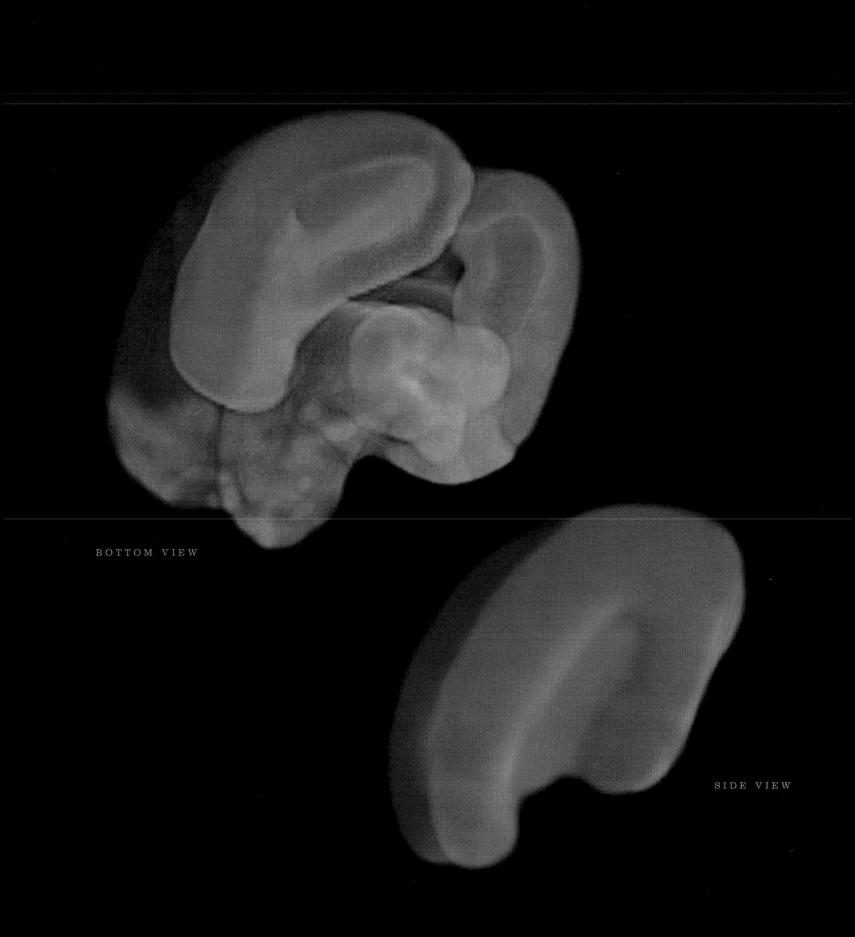

BOTTOM VIEW

SIDE VIEW

As the skeletal system strengthens,
the baby is moving more and
more. The spine consists of about
150 joints and 1,000 ligaments.

ACTING LIKE A BABY

THUMB SUCKING AND NAPPING: As the second trimester draws to a close, the fetus sits fully erect — even more than at birth. This allows the enlarging visceral organs, particularly the liver and heart, to grow and to take up their eventual positions lower down in the body cavity. Some of the intestines are fully retracted into the area known as the pelvic basin.

While the body's proportions begin more nearly to resemble their final form, the more than 200 bones needed to support it in a vertical position continue to harden and knit together. (At birth, the baby has a total of 300 bones, but some later fuse, which is why an adult has just 206.) The human spine is made up of 33 rings, 150 joints, and 1,000 ligaments, which are used to support the body weight. All have begun to form at 6 months, according to Rugh and Shettles.

With little or no fatty deposits underneath, the skin is wrinkled, loose fitting, more delicate than will be needed by the newborn. It's coated with a cheesy secretion that helps keep it supple within the mineralized amniotic fluid and cushions it against abrasions as the fetus twists and turns more actively. By now the fetus dozes and wakes, nestling in favorite positions to sleep, stretching upon arising.

The eyes, which began to develop barely 3 weeks after fertilization as two outward bulges in the most primitive neural tissue, are structurally complete. Now the eyelids begin to flutter and reopen, and the eyes begin to be sensitive to changes in light and dark. The nostrils also are unsealed, and the baby begins to make muscular breathing movements.

With the fetus growing rapidly — to nearly 12 inches in crown-rump length and weighing about one-and-a-half pounds — and moving more actively, protein needs sharply increase. The mother may now be gaining as much as a pound every week.

24 WEEKS

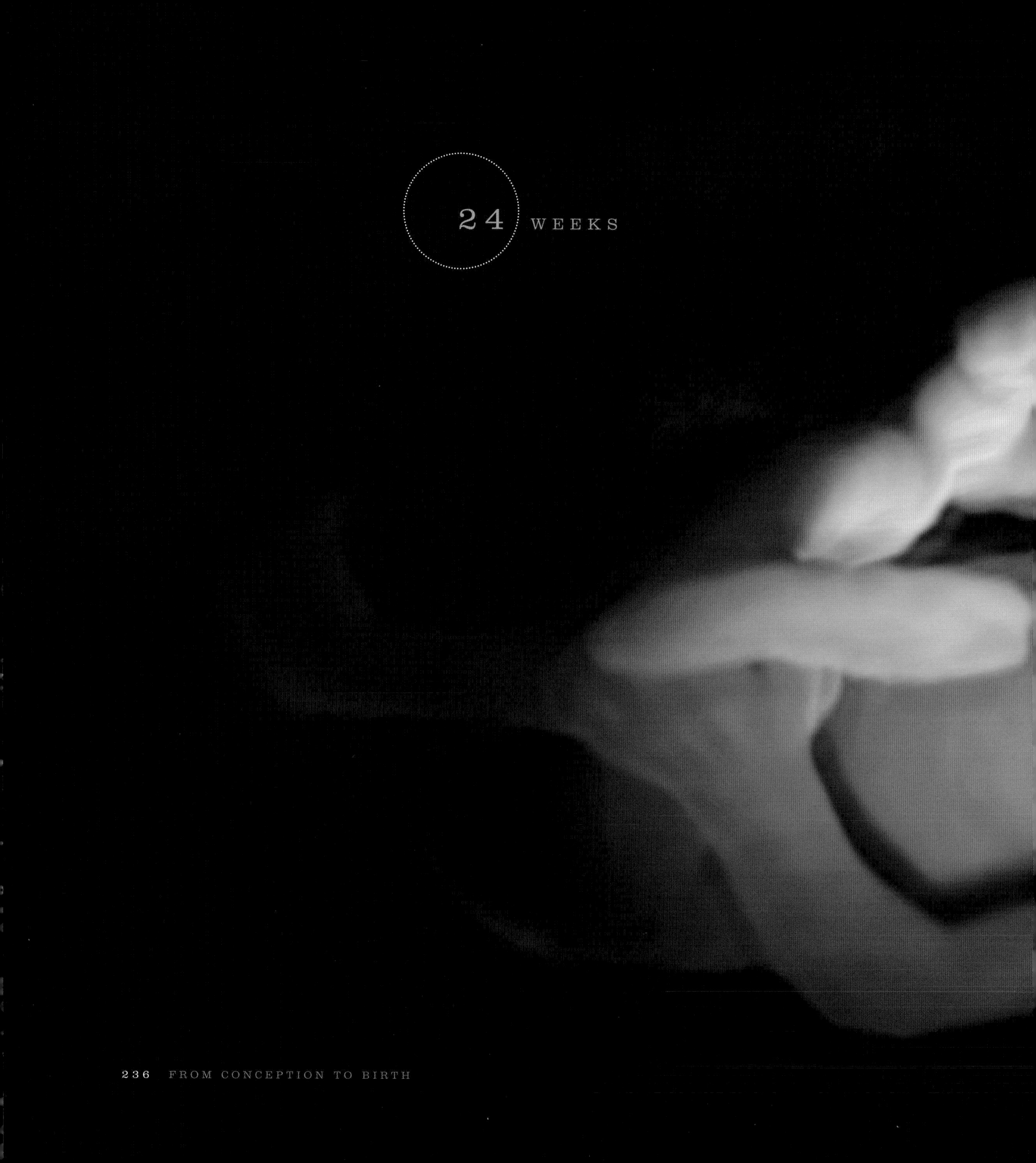

An active mind

Extremely rapid brain growth (which lasts until 5 years after birth) begins. The outlines mimic more of an adult brain shape, but the surface is still smooth.

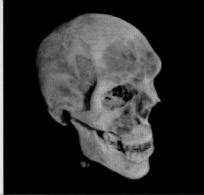

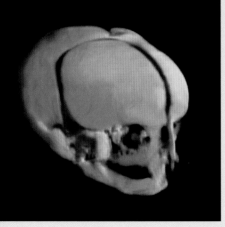

The human skull merges together from pieces. Between them only thin membranes (fontanels) are covering the brain. They are present until the end of the second year, but the skull still develops until we reach adulthood.

➤ Facing page: 14-week fetus. The skeletal system develops from embryonic connective tissue. After 8 weeks the bones start ossifying.

SKELETAL SYSTEM

KNITTING THE BODY TOGETHER: By 6 months, the bony skeleton is fully assembled but not yet connected at the joints. Like the soft wires pressed into the clay of a Gumby figure, the bones support the fetus structurally, allowing twisting and motion. But like a Gumby, it has no mechanical ability — its parts bend, but they can't bend and unbend at will. Your child will need another year at least before the ball-and-socket, hinge, and sliding mechanisms of the hip, knee, and ankle can deliver enough strength and flexibility to permit toddling. In adulthood, the human skeleton is an elegant moving framework, a torqued white Eiffel Tower, levered and double-levered so that (theoretically) the same individual can pirouette, squat-lift a bag of cement, race the 100-meter dash, and perform surgery.

The skull is also synthesized modularly — a helmet molded from 4 partially calcified plates. In utero, the plates are loosely knit rather than fused. This allows the head to taper during delivery, and parents have learned not to be alarmed if their baby's head appears at first elongated or misshapen, or, later, when they discover a soft spot (a "fontanel", French for "little fountain") on the top of the skull. The human brain triples in mass during the first year, and the skull's flexible assembly both protects it and allows it to grow. The skull bones interlock and fuse permanently at about 18 months. Meanwhile, the jawbone continues to protrude, undergoing a final spurt in late adolescence, when parents notice with confused emotions that their children suddenly "look like adults."

Skin

At 25 weeks the skin is wrinkled.

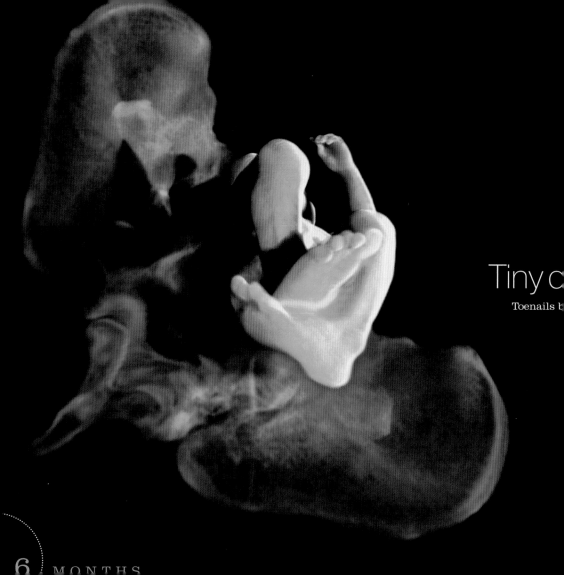

Tiny c

Toenails b

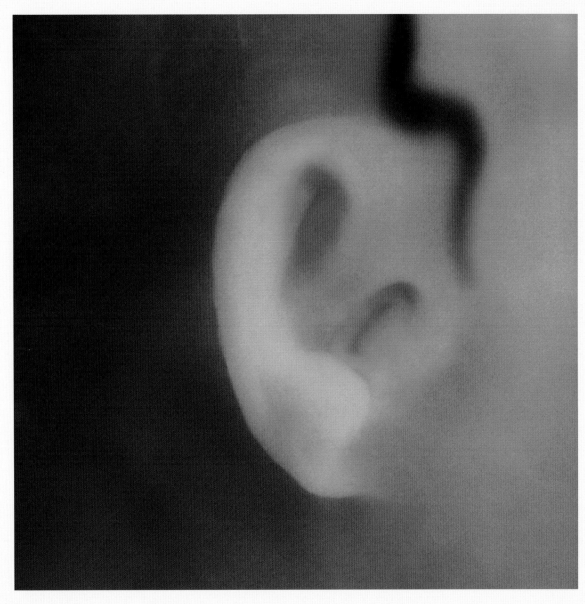

Eyes respond to light, while ears respond to sounds originating outside the womb at 24 weeks.

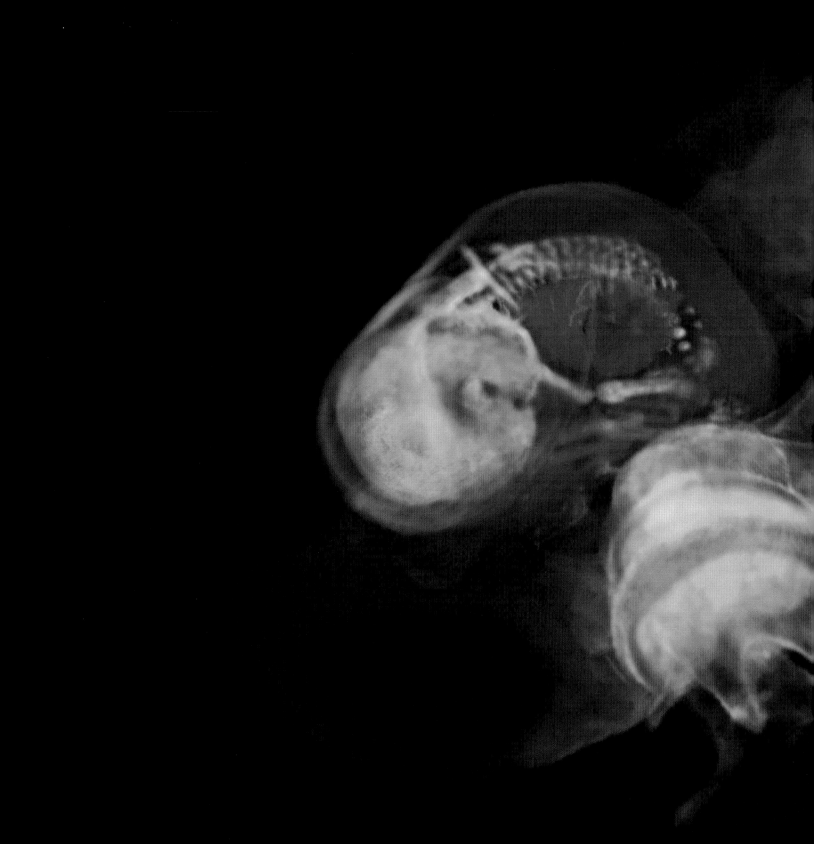

Nestled in

Fetus in female pelvis at 26 weeks. Note the enlarged forebrain. The central nervous system is developed enough to control breathing and body temperature. Lungs are capable of breathing air at 26 weeks. Babies born at 26 weeks or later often survive.

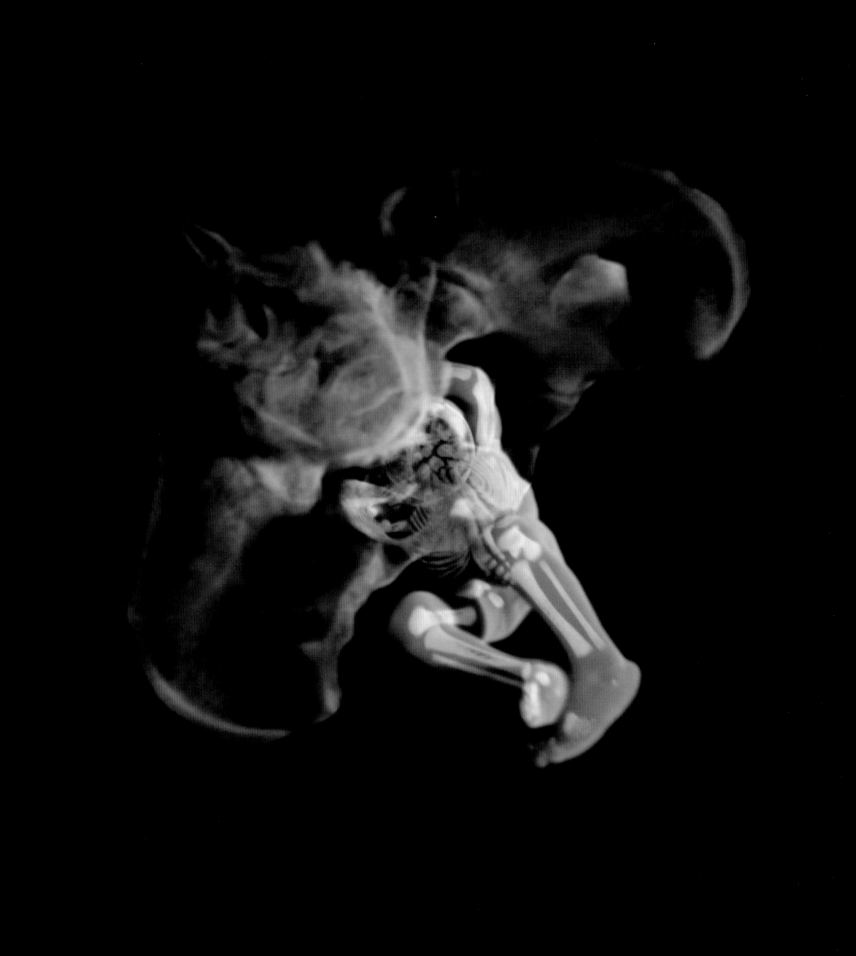

THRESHOLD

VIABILITY: Nowhere in the body is activity enhanced and increased by the emergence of more and more refined forms than in the brain. Up until now the cerebral cortex was relatively unwrinkled, but grooves and furrows have begun bunching on the outer surface, and along with them come greater, more localized functions. In every human, the centers for sight, smell, hearing, speech, walking, and all other sensations and activities arise in precisely the same regions. These are now all in place. The sites and connections governing reason, memory, and imagination remain to be developed, but at this stage the brain has advanced sufficiently to control rhythmic breathing, coordinated contractions in the digestive tract and body temperature—3 major thresholds for living outside the womb. With the lungs now equipped with sufficient air sacs (alveoli) and surrounding blood vessels to allow the baby to exchange oxygen and carbon dioxide, chances for survival outside the mother markedly improve.

Throughout the body, final preparations are underway. Red blood cells are now being produced entirely in the marrow of calcium-reinforced bones. The fine hairy covering known as *lanugo,* which first appeared in the fifth month, disappears almost everywhere except on the back and shoulders, and head hair begins to grow. The skin becomes smoother as fat deposits, which insulate and provide energy, rapidly accumulate. In almost all boys, the testes, which originated inside the body cavity, have descended into the scrotum, paving the way for the future production of viable sperm.

◄ Facing page: Production of red blood cells is entirely taken over by the bone marrow.

In position

The limited space in the uterus forces the legs to be
drawn up in what is known as the fetal position.

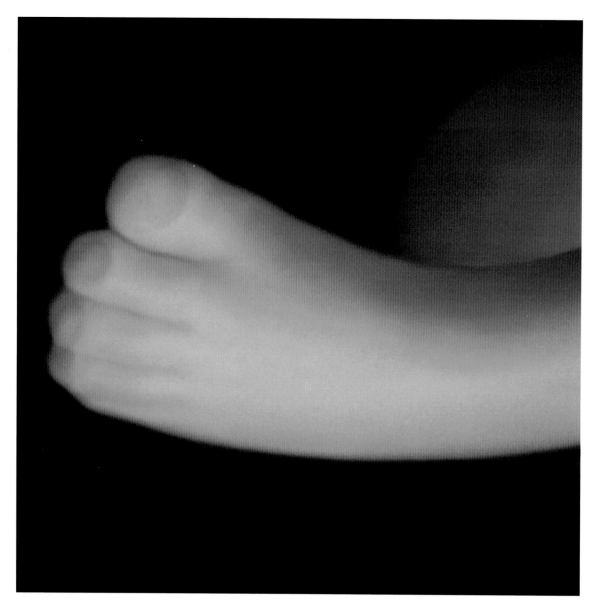

Toenails are fully formed at 30 weeks.

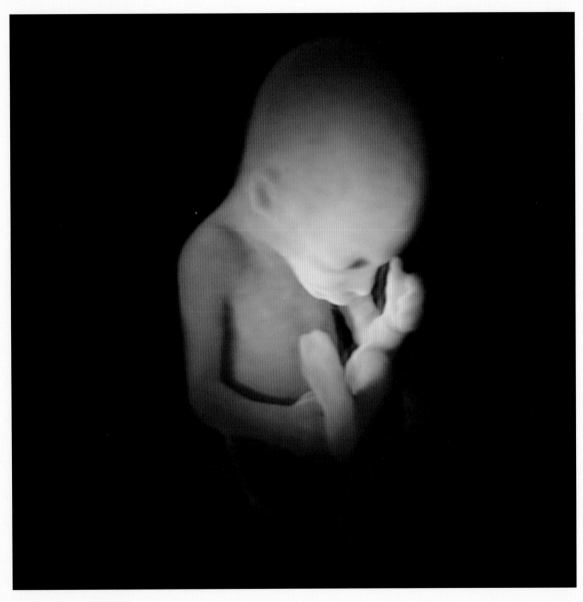

Skin begins to smooth out as fat deposits accumulate underneath. The fat insulates and is an energy source at 28 weeks. The fetus looks paler because of the fat.

CRAMPED LIVING QUARTERS

MOTHER'S HARD WORK: As the baby grows large (crown-rump length 14 inches or more, average weight up to 5 or 6 pounds) the uterus becomes cramped. (Imagine quadruplets.) The legs draw up into the fetal position as, in most cases, the head moves down into the pit of the pelvis. Dimpling at the elbows and knees, creasing at the neck, soft bulges of subcutaneous fat — all give the baby a plump appearance.

The earlier pleasant "water ballet" phase is over, replaced by the mother's feeling gravid, put upon, as if someone were shoving her sharply from the inside. Marie Antoinette reportedly once petitioned her husband, King Louis XVI of France, for royal relief at this point, announcing: "I have come, Sire, to complain of one of your subjects who has been so audacious as to kick me in the belly."

The rate of weight gain in both fetus and mother begins to slow in anticipation of birth. (If the fetus continued to grow at the same rate as previously, if would weigh an estimated 200 pounds by its first birthday.) Still, because the fetal digestive track — stomach, liver, pancreas, intestines — remains immature, the baby stores nutrition taken from the mother against the risk of early birth. Meconium — dark green mucous and dead cells from the liver, pancreas, and gall bladder — forms in the intestines.

All sense organs are fully functional, and the baby looks much as it will upon arrival, although most now have blue eyes regardless of permanent color. Final formation of eye pigmentation requires exposure to light and usually happens a few weeks after delivery.

Right: Babies have their own immune system 6 weeks before birth.

8 MONTHS

Racing mind

The brain still grows rapidly as growth
of the body slows down.

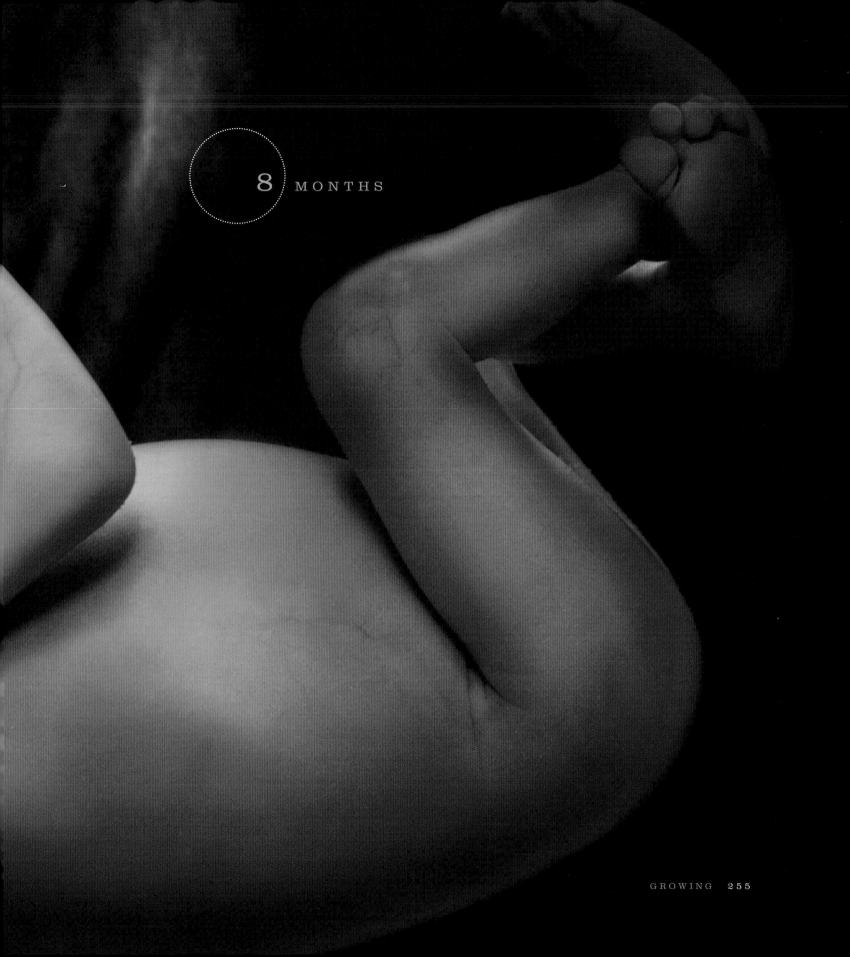

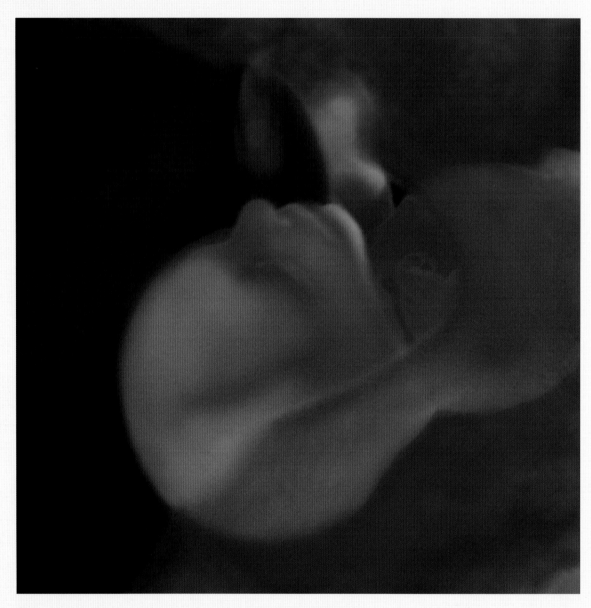

Skin appears light pink because of blood vessels close to its surface.

Fingernails reach over fingertips and the fetus can scratch itself from week 32.

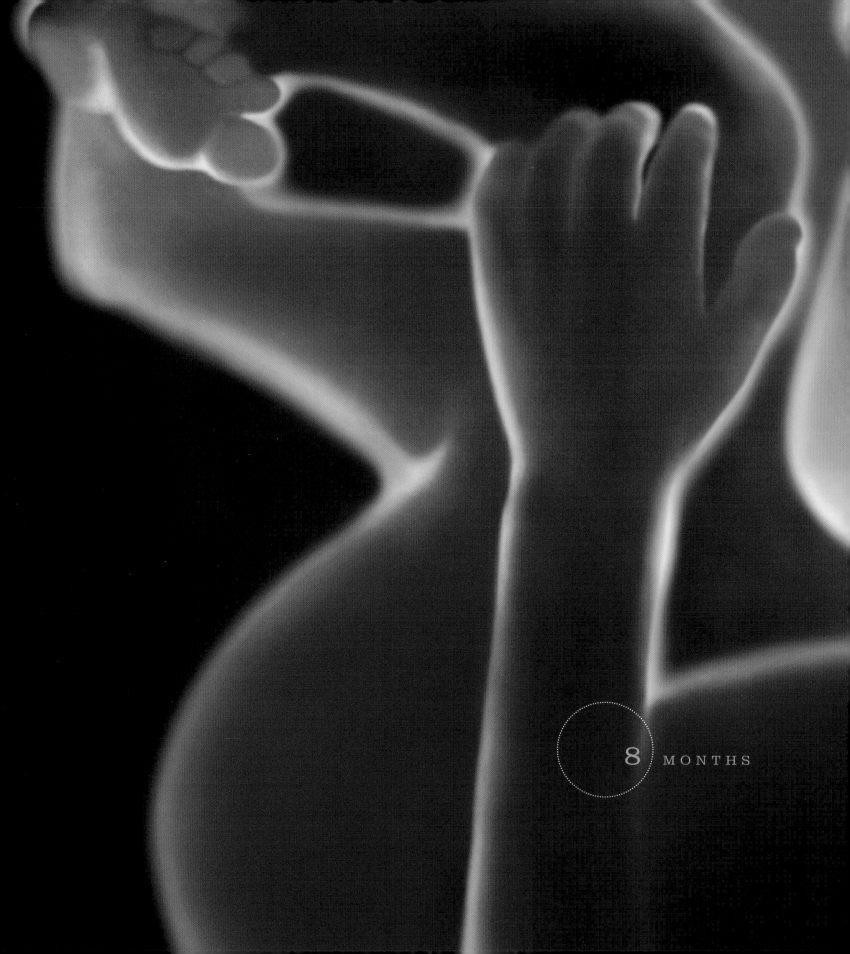

8 MONTHS

Baby fat

Body is round and plump due to new fat
storage, keeping the body temperature at about
32 F° above maternal temperature.

Fetal abdomen is large and round
mainly due to the liver, which is
producing red blood cells. Space
becomes too small. The baby is
ready to be born.

READY FOR THE WORLD

MOTHER'S FINAL GIFTS: The baby may seem less active during this month, but only because there is so little room for activity. Girls typically weigh between 6 and 8 pounds and measure 14 to 15 inches crown to rump (19 to 20 inches overall). Boys may be slightly larger. Your baby could be delivered at any time, although final touches remain underway; indeed, many features continue developing after birth and others throughout life.

Most notable perhaps is the skull, which isn't yet fully solid but contains 5 unconnected bony plates called *fontanels* (little fountains). A safety precaution for reducing the skull's diameter for easier passage through the birth canal, this collapsible framework suits both the fetal brain's need for protection and the mother's need to deliver her child. The fontanels allow the head to elongate and mold during childbirth, then return to a rounded shape.

Also significant is the transfer of immunities from mother to child. During this final month, the baby usually receives a temporary supply of antibodies from its mother's blood. These disease-resisting proteins will protect the baby against such childhood diseases as mumps, measles, whooping cough, and scarlet fever, as well as some bacterial strains and even cold and flu viruses, until the child can formulate its own.

By late in the ninth month, fingernails and toenails may have grown so long that the baby has suffered scratches. The gums are ridged, in some cases giving the false impression of emerging teeth. From the original heap of soap-bubble-like cells that wafted down from the fallopian tube 9 months earlier and attached to the uterine wall has grown a complete human being, optimized for survival in every detail.

Birth

A NEW VOICE IN THE WORLD

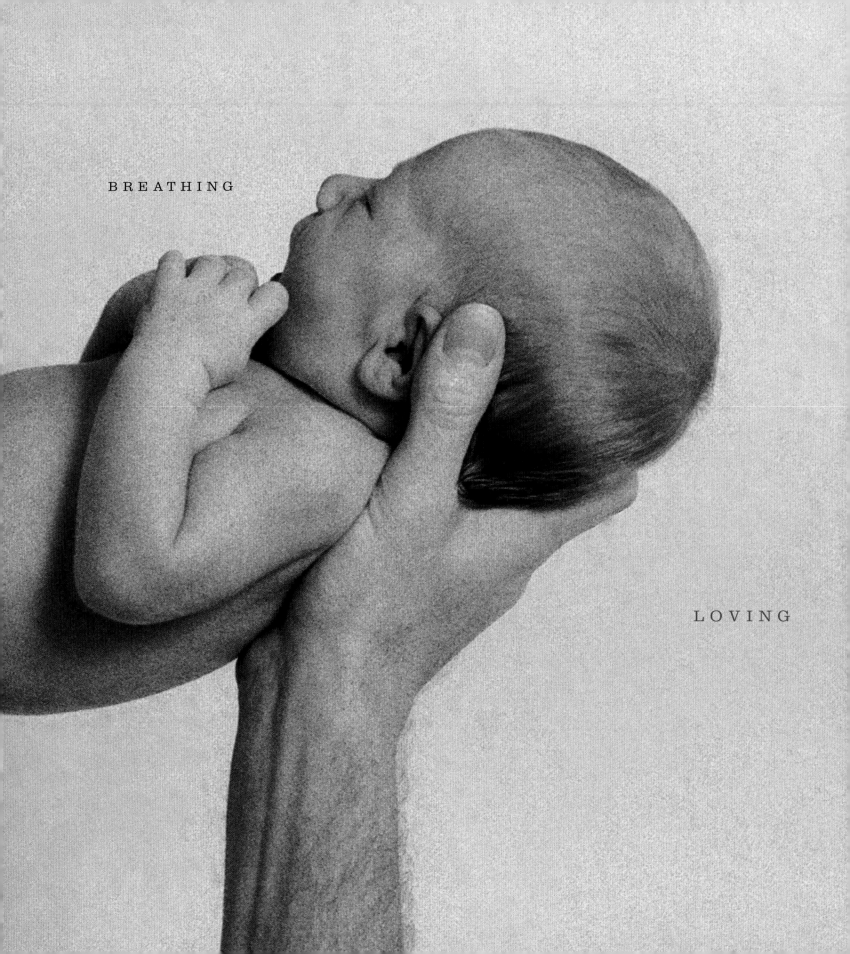

BREATHING

LOVING

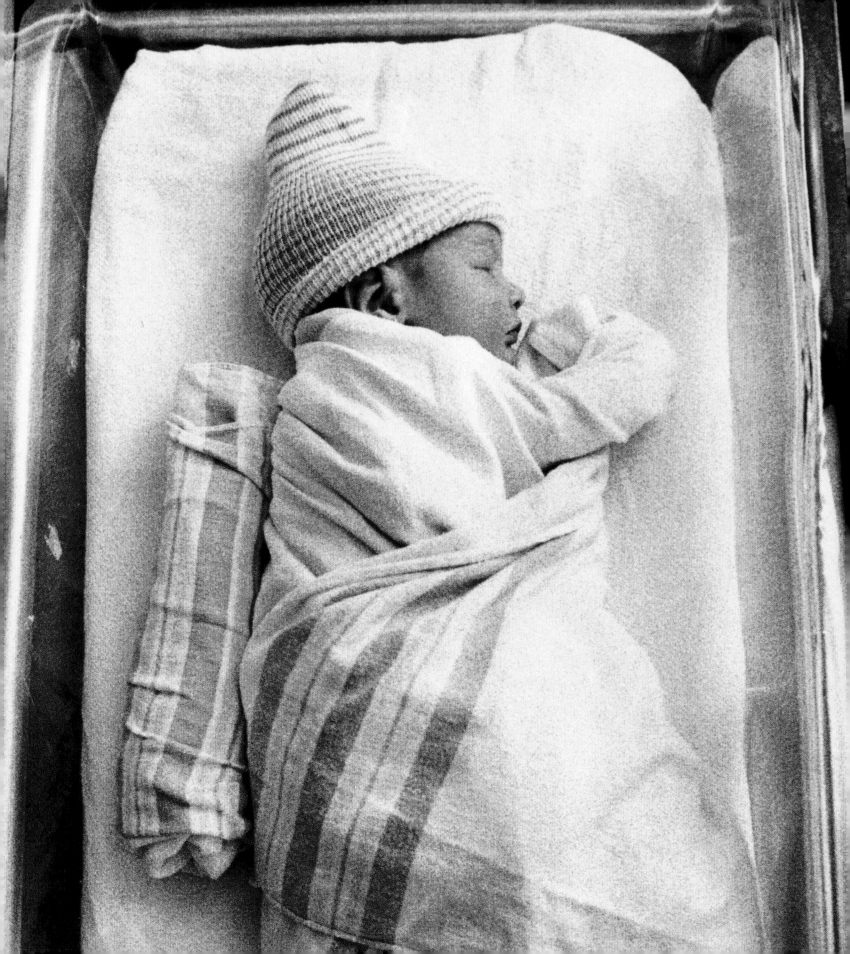

05

Birth

The baby's advent into the world marks epic changes for newborn and mother alike. Major fetal organs like the lungs, stomach, intestines, and kidneys, which have remained untested because there was no oxygen to breathe, food to digest, or wastes to eliminate, all must start functioning at once — in some cases within seconds of delivery. Others already in use like the heart and brain must adapt without hesitation to a world for which they have been meticulously designed and prepared but that differs utterly from the one in which they've functioned. Blood coming into the heart must suddenly be shunted to the lungs, meaning the opening between the 2 upper chambers (necessary for circulation when oxygen enters the bloodstream through the placenta) needs to be sealed at once. The nervous system immediately starts making use of information from eyes practically blinded by a blizzard of incoming data (not to mention the eyedrops doctors give them to prevent infections). Returning to the idea that we are mechanical buildings, it is as if we were constructed under the ocean, then

thrust up onto dry land in perfect working order —
and immediately pressed into service.

A baby, unlike a fetus, must sustain itself
against shocks, changes in temperature, disease; still
later, against the rough and tumble of school, work,
dating, loss, and having children of one's own.
Starting at birth, it must negotiate for itself. Thus the
days of being a parasitic dependent on an acquiescent host are (one hopes) over.

For the mother, there is a corresponding loss
of primacy in the fetus' life. She has been everything — food, climate, shelter, security — literally the air
her child has breathed. She was occupied by an
invader, one who caused her willingly to put up with
enormous discomfort and changes in herself, who is
now outside her, crying out.

The "sweet sorrow" of parting may be lost on
her as the shock of separation sets in. But babies,
like fetuses, have a remarkable way of insinuating
themselves into their parents' lives. Indeed, their
helpless perfection and helplessness gives them the
power, as every proud parent learns, to attract not
only admirers and sympathizers, but adoring crowds.

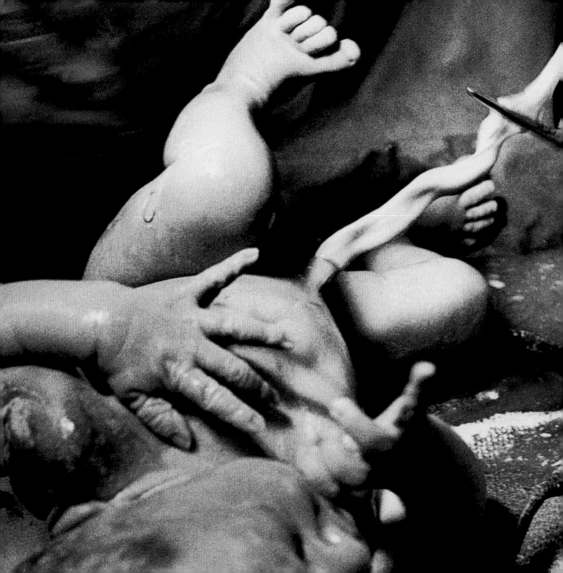

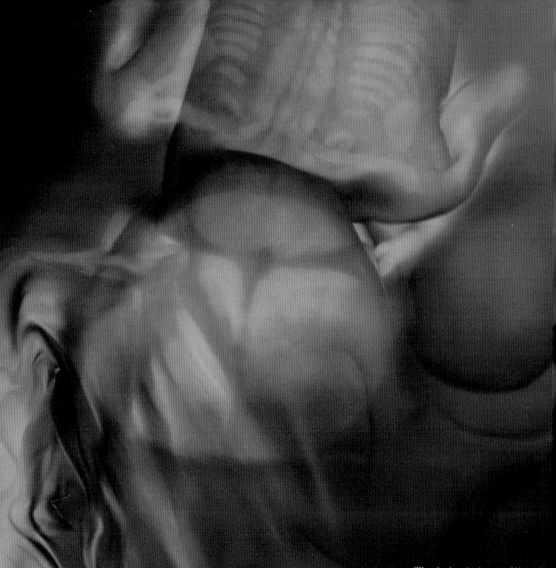

The baby is in position t

LABOR

AT LONG LAST: Since the first few days after conception, the mission of the uterus has been to confine and protect the developing fetus; now its job is to expel the fetus and restore itself for future pregnancies. The body's handmaidens for achieving physiological role reversals are hormones, and to trigger the practical and mechanical job of delivering the fetus from the womb, the mother's brain begins to release oxytocin into the bloodstream. The so-called "love hormone," oxytocin jump-starts the process of labor by stimulating the uterine muscles to contract.

Another hormone secreted by the ovary relaxes the cervix, which, stretched by the baby's head, prepares now to widen gradually from half a centimeter — about the size of a peanut — to 10 centimeters in diameter. A plug of mucous in the cervix dislodges and passes out the vagina, often slightly mixed with blood. Since early contractions are normally so mild and short that some women scarcely notice them, this "bloody show" may be the first sign that labor has begun. Meanwhile, the vagina secretes a complex carbohydrate called glycogen, which metabolizes on the spot to lactic acid, an infection-fighting agent.

As contractions occur more frequently (3 to 5 minutes apart), they also become stronger and longer-lasting. An expectant mother may be able to keep going about her business, but these more frequent contractions generally cause greater and greater discomfort (hence "labor pains") and are a signal that the time for delivery is impending.

As Natalie Angier and others have noted, parturition (giving birth) is a social process. Unlike other primates, humans are not adapted to give birth alone. Our upright posture, and the largeness of the baby's head, change the mechanics of delivery, making human childbirth relatively painful and prolonged. Most newborns face backward as they emerge from the vagina, so the mother would risk damaging the baby's head and neck if she tried to assist the passage with her hands. The point: However exclusive the mother-fetus may have become during pregnancy, other people are now necessary, can be a great help and comfort, and should be selected with care.

C-SECTION

1 IN 7 : Although most expectant couples assume their child will be born vaginally, 1 in 7 newborns in the United States is delivered through an incision in the mother's abdomen. Most commonly, labor stalls because the baby's head is too large to fit through the pelvis, or else the fetus or mother experiences distress. In either case, the risk of allowing natural childbirth to continue becomes unacceptable, and the child is delivered promptly through surgery. Caesarean sections may also be called for in cases where the placenta is abnormally low and covers the cervical opening, or if the baby is in the breech position and cannot be turned. A C-section is a major operation, performed under anesthesia, and recovery time for the mother is longer than for a vaginal birth.

The rule of thumb used to be: Once a C-section, always a C-section. That practice no longer holds. If the reason why a Caesarean was done in the first place doesn't recur, a mother now has a 60 percent chance of subsequently delivering vaginally — so called VBAC (vaginal birth after Caesarean).

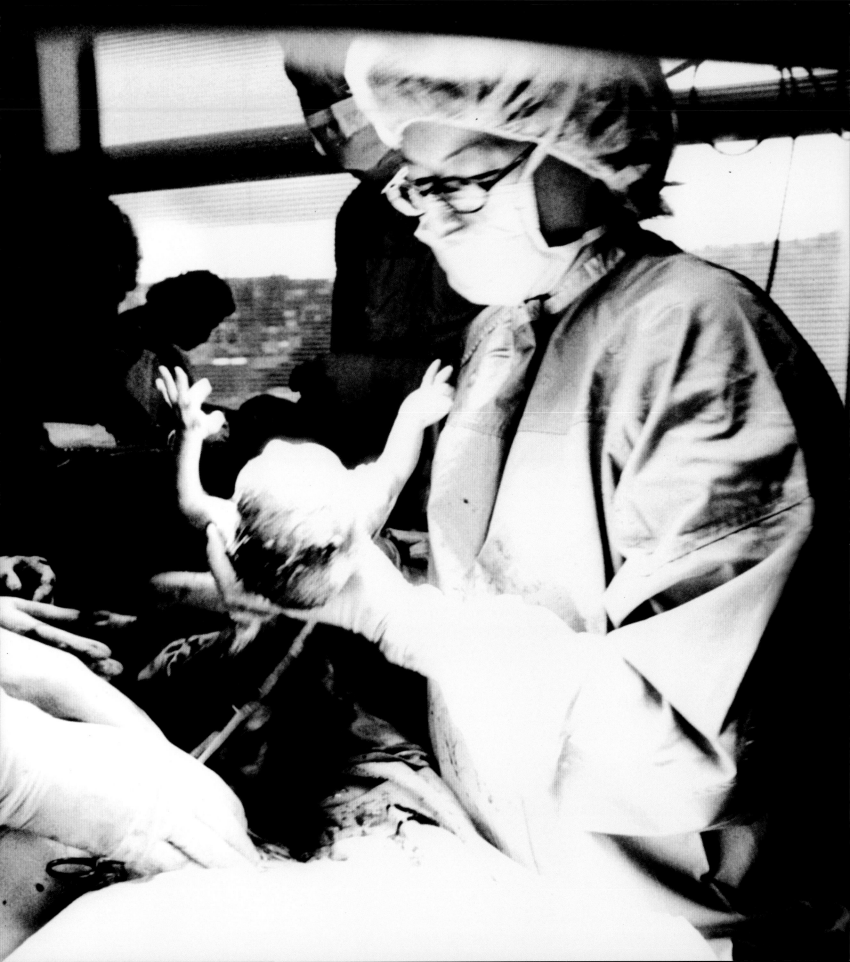

1

IN POSITION

Uterine contractions felt throughout the last weeks
of the pregnancy. The uterus becomes really stiff
during these periods that we call "labor pains."

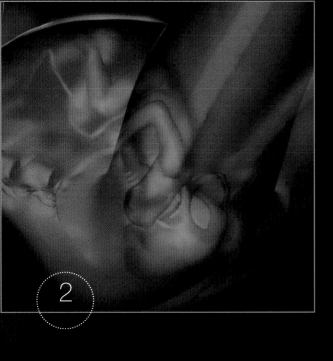

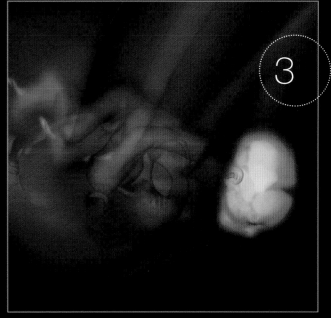

UTERUS OPENS

The baby rotates down; the cervix dilates; the baby pushes out with the contractions of the mother.

BABY EMERGES

In most deliveries, the baby comes out headfirst and face-down.

Head first

Humans have an extremely oversized head compared to their body. In order to squeeze the baby through the pelvic hole, the mother's bone has to pop open in the middle. The first time is always the hardest.

First moments

The baby has been put through a traumatic experience.
The still large adrenalin glands release a large amount
of stress hormones to make the event and the oxygen
deficiency bearable.

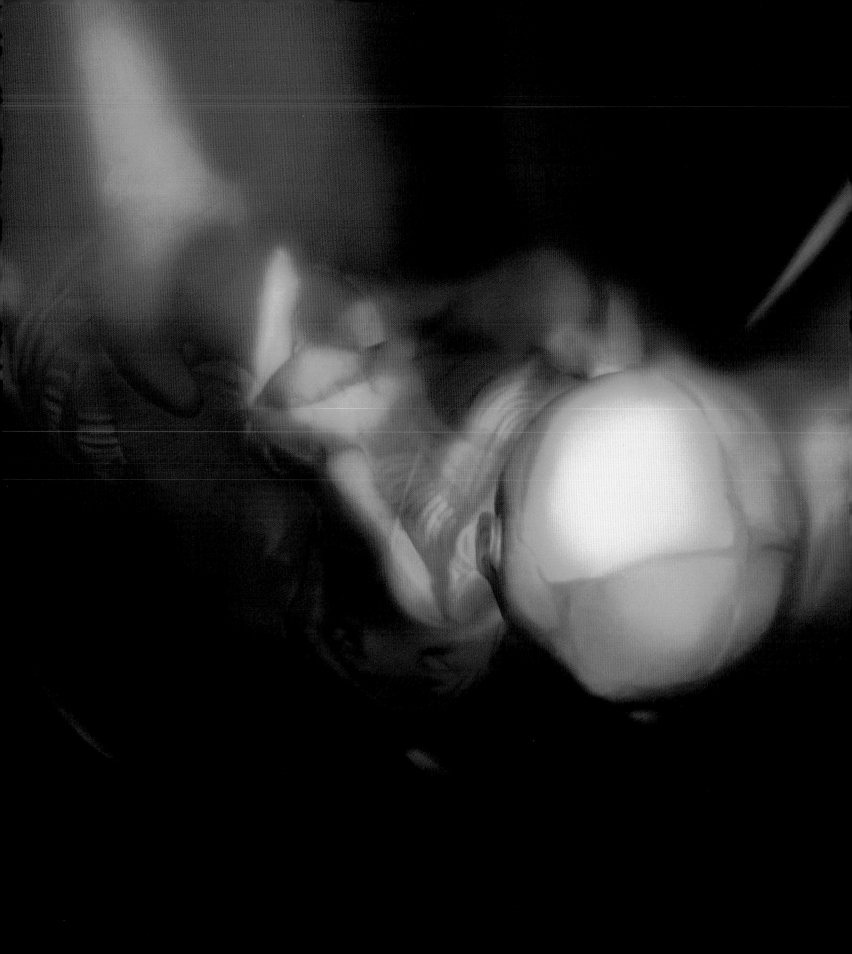

DELIVERY

WELCOME TO THE WORLD: Once the cervix is fully dilated, birth begins. The mother's contractions—coming every 2 to 3 minutes and lasting about a minute—so far have been involuntary. But now she is free to help by bearing down and pushing hard with her abdominal muscles. Early on during this stage, the amniotic sac ruptures (or may be surgically ruptured by the obstetrician to help speed up delivery). Amniotic fluid gushes from the vagina, leaving the fetus suddenly unprotected. With the fetus now subject to germs, adhesions, and changes in temperature, aseptic conditions and prompt delivery are crucial to newborn health and vitality.

In almost all cases (3.5 percent of babies are born feet first—"breeches down," or simply breech) the baby's head pushes through the fully dilated cervix during contractions and appears as the opening of the vulva widens. Then it recedes between contractions. Once the head fully appears (so-called "crowning"), doctors, nurses, and midwives are trained to grip the baby's head and help pull it through. Because the fontanels have not yet fused, the baby's head is pliable and may emerge misshapen—a temporary condition.

Once the baby has emerged, the respiratory apparatus kicks swiftly into action. In confinement, the fetus's lungs remained either collapsed or partially filled with amniotic fluid, its millions of air sacs compressed, soaked, and untested. At birth, after the umbilical cord is tied and cut following a last surge of blood from the placenta, the baby's oxygen supply is abruptly terminated. As circulation continues in the baby's body, carbon dioxide builds up in the bloodstream, causing the respiratory control center in the brain stem to signal the muscles of the diaphragm and rib cage to begin the regulated pattern of movement first rehearsed months earlier. The muscles contract, expanding the rib cage, pulling the diaphragm downward, and allowing room for the lungs to fill like sails in a gale.

Upon this first great inhalation—and the vigorous exhalation that follows—the baby cries out. No other behavior indicates more positively that the baby has adjusted to life outside the womb, and parents and others can be forgiven their tremendous relief and excitement if they cry out too.

ACKNOWLEDGMENTS

Bill Thomas, Editor in Chief, Doubleday
I want to compliment Bill for his ability to see the potential of this book when so few could. I also want to thank him for his aesthetic and managerial insights and for making this entire process one of the most fulfilling professional experiences of my career.

Attila Ambrus, Senior Producer, Anatomical Travelogue, Inc.
The hours and excellence he devoted to the production of this book exceeded my every expectation. His aesthetic contributions and his technical excellence have elevated this book to a level that has made all those involved very proud.

My family: Susan and Andreas
I thank them for their immeasurable love, support, patience, innocence, and laughter. I only hope that I can give them back even a portion of what they have given me.

Anatomical Travelogue participants:
Glenn Ball, Laszlo Balogh, Chad Capeland, Diana Diriwaechter, Ildiko Fodor, Joanne Sommers Handley, Wan Seob Lee, Benjamin Lipman, Mark Mazaitis, Elias Papatheodorou, Kwang Park, Levente Szileszky.

E. C. Lockett, Imaging Specialist, the National Museum of Health and Medicine of the Armed Forces Institute of Pathology.

William Discher, Imaging Specialist, the National Museum of Health and Medicine of the Armed Forces Institute of Pathology.

Center for In Vivo Microscopy, Department of Radiology, Duke University Medical Center.

Biomedical Magnetic Resonance Laboratory, University of Illinois, College of Medicine at Urbana-Champaign.

Olympus America, Inc., Olympus Scientific Equipment Group.

Marlin Minks, Former Creative Director, Anatomical Travelogue.

Carlo Bulletti, M.D., Director of Physiopathology of Reproduction, Rimini's General Hospital, University of Bologna and Health Unit of Rimini, Professor of Biotechnology of Reproduction in the School for Biotechnology at the Faculty of Medicine of the University of Bologna, Italy; Adjunct Associate Professor of Obstetrics and Gynecology at New York University.

P. Schwartz and H. W. Michelmann, University of Goettingen, Germany.

The National Institute of Child Health and Human Development (NICHD), National Institute of Health.

National Institute of Health.

Douglas T. Carrell, Ph.D., H.C.L.D., Associate Professor of Surgery, Obstetrics, and Gynecology, and Physiology, Director of

IVF and Andrology Laboratories; and Benjamin R. Emery, B.S., University of Utah School of Medicine, Salt Lake City, Utah.

Raymond F. Gasser, Ph.D., Senior Professor of Human Embryology, Department of Cell Biology and Anatomy, Louisiana State University Health Science Center, New Orleans, Louisiana.

Lewis C. Krey, Ph.D., Anna Blaszczyk, M.S., Caroline McCaffrey, Ph.D., and Alexis Adler, B.S. Program for In Vitro Fertilization, Reproductive Surgery, and Infertility, Department of Obstetrics and Gynecology, New York University School of Medicine .

Stark Design: Adriane Stark and Craig Bailey
You made a very difficult and complicated process very worth it. Thanks for the beauty of your design, and your consummate professionalism.

Barry Werth, writer: "Your text is Mozartian," Steve Rubin, Publisher of Doubleday.

Kendra Harpster, Assistant Editor, Doubleday.

Rebecca Holland, Publishing Manager, Doubleday.

Kim Cacho, Doubleday Production.

Jennifer Rudolph Walsh, Senior Vice President, Head of Literary Department, William Morris Agency.

All images were produced on HP Workstations.

Scientific visualization and volumes software developed in collaboration with Volume Graphics GmbH, Germany (www.volumegraphics.com).

PHOTOGRAPHY CREDITS

Page i, © Barnaby Hall; page viii, © Sandra Wavrick / Photonica; page xviii, © Michael Kelley / Getty Images; pages 2-9, © Barnaby Hall; pages 12, 13, © Jane Yeomans / Photonica; page 37, © Andreas Pollok / Getty Images; pages 46, 47, 76, 77, © Jane Yeomans; pages 196, 197, © Erik Rank / Photonica; pages 262, 263, © Owen Franken / Corbis Images; page 264, © Armen Kachaturian / Photonica; page 267, © Elke Hesser / Photonica; pages 270, 271, © Spike / Getty Images; page 279, © David Roth / Getty Images; pages 284, 285, © Jane Yeomans / Photonica.

We also wish to thank: E. C. Lockett and William Discher, Imaging Specialists, the National Museum of Health and Medicine of the Armed Forces Institute of Pathology; Marlin Minks, Former Creative Director, Anatomical Travelogue; P. Schwartz and H. W. Michelmann, University of Goettingen, Germany, electron microscopy; Douglas T. Carrell, Ph.D., H.C.L.D., Associate Professor of Surgery, Obstetrics, and Gynecology, and Physiology, Director of IVF and Andrology Laboratories; Benjamin R. Emery, B.S., University of Utah School of Medicine, Salt Lake City, Utah; National Museum of Health and Medicine/Carnegie Human Embryology Collection. Lewis C. Krey, Ph.D., Anna Blaszczyk, M.S., Caroline McCaffrey, Ph.D., and Alexis Adler, B.S., Program for In Vitro Fertilization, Reproductive Surgery, and Infertility, Department of Obstetrics and Gynecology, New York University School of Medicine.

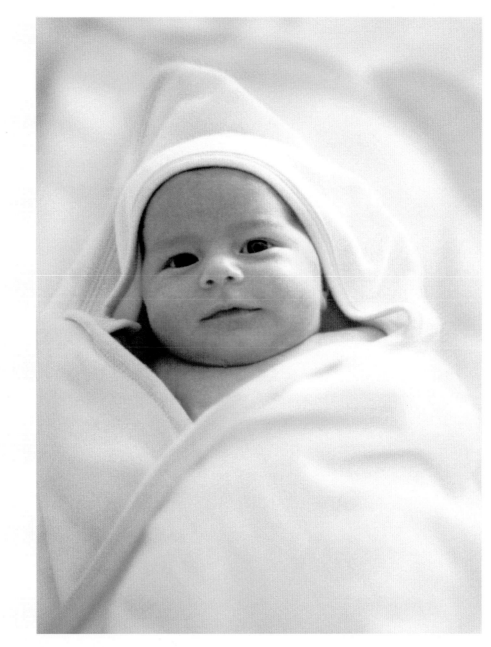

The author's son, Andreas